CafeOBJ Report

The Language, Proof Techniques, and Methodologies
for Object-Oriented Algebraic Specification

AMAST SERIES IN COMPUTING

AMAST Series in Computing: Vol. 6

CafeOBJ Report

The Language, Proof Techniques, and Methodologies
for Object-Oriented Algebraic Specification

Răzvan Diaconescu
Kokichi Futatsugi

Japan Advanced Institute of Science and Technology

World Scientific
Singapore • New Jersey • London • Hong Kong

Published by

World Scientific Publishing Co. Pte. Ltd.

P O Box 128, Farrer Road, Singapore 912805

USA office: Suite 1B, 1060 Main Street, River Edge, NJ 07661

UK office: 57 Shelton Street, Covent Garden, London WC2H 9HE

British Library Cataloguing-in-Publication Data
A catalogue record for this book is available from the British Library.

CAFEOBJ REPORT

ISBN 981-02-3513-5

This book is printed on acid-free paper.

Printed in Singapore by Uto-Print

Preface

This is a report on the formal definition of the CafeOBJ algebraic specification language, which is a modern successor of the famous algebraic language OBJ. While the equational core of CafeOBJ is just a re-shaping of OBJ, CafeOBJ significantly extends OBJ by incorporating several recent major developments in the area of algebraic specification, such as behavioural specification and rewriting logic.

The definition of the language parallels its logical semantics based on the so-called *institutions* which also provide a methodological framework for structuring the presentation of the basic constructs of the language and their semantics.

This Report presents all basic constructs of the language together with their semantics and addressing both the programming in-the-small and in-the-large levels, but also discusses proof systems and technologies, as well as methodologies. Examples are provided all over the Report as intuitive support for the definitions of the constructs and for illustrating proof techniques and methodologies.

Purpose

This Report presents the formal definition of CafeOBJ independently of particular implementations of the language. Therefore it serves multiple purposes:

- as a reference document for the language,
- as an introduction to the language,

- as a support for implementations of the language.

Structure

Part I (Section 1–5) deals with *basic specifications* and their proof systems, discussing the various kinds of specifications classified by the underlying paradigm; Part II (Sections 6–9) deals with *structuring specifications* discussing the module system and libraries; Part III discusses techniques for proving properties of specifications; and Part IV discusses methodologies in CafeOBJ. The underlying semantics of each of the language constructs is provided along with its definition. Finally, the appendix provides a complete grammar for the CafeOBJ syntax.

Many examples are prepared and scattered all over the book to make the concepts presented be understood in a concrete way. We recommend readers to make use of these examples extensively. For example, trying to read through all the examples with minimum necessary materials other than examples is a good way to understand how to write specifications in CafeOBJ.

Related documents

This Report should be regarded as sitting in between two other important documents: the *Logical Semantics* document [15], and the (various) *User's Manuals*. The CafeOBJ Report is more concrete than the former which gives an open and abstract institution-based semantics for CafeOBJ and points in a structured way to all relevant theoretical developments underlying the semantics of the language, but without addressing all the details of the language constructs. In fact, the CafeOBJ Report can be regarded as just explicitating the Logical Semantics of the language.

On the other hand, CafeOBJ Report is providing formal support for user's manuals corresponding to various possible CafeOBJ implementations, such an example being the SRA implementation of CafeOBJ [51]. While the SRA implementation comes very close to the full definition of CafeOBJ which is given in this Report, some gaps still remain.

Other documents regarding CafeOBJ, such as scientific papers or reports, can be obtained from the CafeOBJ World Wide Web home page at `http://www.ldl.jaist.ac.jp/cafeobj/`.

Acknowledgments

The authors wish to thank Joseph Goguen for his permanent encouragement and technical help during the CafeOBJ project; several major design decisions for CafeOBJ were made after consultations with him. We thank MITI/IPA for supporting the CafeOBJ project, thus making the fundamental implementation of CafeOBJ language system and the writing of this book possible. We are also grateful to all members of the Language Design Laboratory at JAIST who provided an excellent team, and to all members of the CafeOBJ project for the basic working environment within the project. We are especially grateful to Shusaku Iida for close collaboration on applications and for help with preparing the final version of the manuscript. Special thanks deserve the members of the SRA team (Toshimi Sawada, Ataru Nakagawa, and Makoto Ishisone) who implemented the system; without their close collaboration this work would not have been possible. We thank Dorel Lucanu for our collaboration in the area of examples, Virgil Căzănescu for his deep technical comments and for spotting several subtle technical bugs on earlier drafts of this Report, Martin Wirsing who, as AMAST referee, read this Report in great detail and made several valuable suggestions for improving the presentation, and Teodor Rus for his great support for publishing this book.

R.D. and K.F.
Tatsunokuchi
Feb., 1998

Contents

Overview of CafeOBJ

A Brief History

CafeOBJ is a direct successor of OBJ, maybe the most famous algebraic language.

The origins of OBJ can be traced back to Goguen's gradual realization, around 1970, that Lawvere's characterization of the natural numbers as a certain initial algebra [46] could be extended to other data structures of interest for Computing Science. The influence of Saunders Mac Lane was also important during that period, leading to the beginning of the famous ADJ group [27] led by Goguen. During the ADJ group period, a mathematical theory of abstract data types as initial algebras was developed. Together with considering term rewriting as the computational side of abstract data types, this constitutes the pillar of the OBJ basic specifications level. It is important to mention that from the very beginning the design of OBJ had been emerging directly from clean and elegant mathematical theories, this process being (rather subtle) reflected as one of the main strength of the language.

Another major step in the development of OBJ was the relativization of algebraic specification over any logic due to Goguen and Burstall's *institutions* (categorical abstract model theory for specification and programming). This pushed the theory of algebraic specification into a modern age. At the beginning institutions provided support for developing advanced structuring specification techniques (i.e., module composition systems) independently of the actual formalism, as emerging from the research on Clear [5]. However, today, after nearly two decades, their significance has been widely expanded. For example, institutions support in an essential way the design of multi-paradigm (declarative) systems.

It could be said that initial algebra semantics, rewriting, and institutions, are the conceptual pillars of the OBJ tradition. Their development and refinement can be easily noticed if you take a close look at the chain of successive versions of OBJ, culminating with the current OBJ [22, 40].

Following the vitally important idea of module composition of Clear, several attempts of implementing modularized algebraic specification languages were done including the early pioneering design and implementation of HISP language [24, 25, 20]. After these experiences, the stabilization of the OBJ design (and its most prominent implementation at SRI) started after the design and prototype implementation of OBJ2 at SRI in 1984 [22].[1] It coincides with several attempts to extend OBJ towards other paradigms, most notably constraint logic programming [35, 8], object-oriented programming [37], and generic theorem proving [56]. Although, due to the indisputable strength of algebraic specification, all these attempts were successful, the interest of the OBJ community has been recently shifting towards a new generation of OBJ focusing more on the recent internal developments in algebraic specification rather than in integrating powerful paradigms from the outside world. Two such examples are CafeOBJ [26, 15, 23] and Maude [50, 47]. With respect to CafeOBJ, although some experimental design and implementation were done in the past, this Report is the first definitive definition of the language.

Many methodological works were done around OBJ and CafeOBJ. We refer here to [21] for a general account for understanding the potentiality of algebraic techniques in software engineering, to [18, 19] for parameterized programming, to [53, 54] for communication protocol specifications, to [43] for distributed system specifications, and to [52] for object-oriented specifications. These works can be seen as a sort of precursor for the design and definition of CafeOBJ in this Report.

---

[1] A comprehensive survey of the algebraic specification formalisms of 1980's, including all major languages of that period can be found in [60].

Paradigms and Features

This section gives a brief overview of the main features of CafeOBJ, all of them reflected in the logical semantics. These should be understood in their combination rather than as separated features. Combining some of these features (sometimes all of them!) results in new specification/programming paradigms that are often more powerful than the simple sum of the paradigms corresponding to the individual features, an example being given by behavioural rewriting specification [13].

Equational specification and programming

This is inherited from OBJ [40, 22] and constitutes the basis of the language, the other features being built on top of it. As with OBJ, CafeOBJ is *executable*, which gives an elegant declarative way of functional programming, often referred as *algebraic programming*.

Rewriting logic specification

This is based on a simplified version of Meseguer's *rewriting logic* [49] (abbreviated **RWL**) specification framework for concurrent system which gives a non-trivial extension of traditional algebraic specification towards concurrency. RWL incorporates many different models of concurrency in a natural, simple, and elegant way, thus giving CafeOBJ a wide range of applications. Unlike Maude [47], the current CafeOBJ design does not fully support *labeled* RWL which permits full reasoning about multiple transitions between states (or system configurations), but provides proof support for reasoning about the *existence* of transitions between states (or configurations) of concurrent systems via a built-in predicate with dynamic definition encoding both the proof theory of RWL and the user defined transitions (rules) into equational logic.

However, from a methodological perspective, CafeOBJ develops the use of RWL transitions for *declarative encoding of algorithms* (see Section 18) as well as for specifying and verifying transition systems.

Behavioural concurrent specification

Behavioural specification [31, 33] provides another novel generalization of traditional algebraic specification but in a different direction. Behavioural specification characterize how objects (and systems) *behave*, not how they are implemented. This new form of abstraction can be very powerful in the specification or verification of software systems since it naturally embeds other useful paradigms such as concurrency, object-orientation, constraints, nondeterminism, etc. (see [33] for details). Behavioural abstraction is achieved by using specification with hidden sorts and a behavioural concept of satisfaction based on the idea of indistiguishability of states that are observationally the same, which also generalizes process algebra and transition systems (see [33]).

CafeOBJ directly supports behavioural specification and its proof theory through special language constructs, such as hidden sorts, behavioural operations and behavioural axioms (stating behavioural satisfaction). The advanced coinduction proof method receives some support in CafeOBJ via a default coinduction relation. In CafeOBJ coinduction can be used either in the classical HSA sense [33] for proving behavioural equivalence of states of objects, or for proving behavioural transitions (which appear when applying behavioural abstraction to RWL).

Object orientation

In CafeOBJ there are (at least) two sources of object-orientation. The first is given by the rewriting logic á la Maude treatment of objects which is specific data (i.e., record) oriented, the second one is given by the behavioural specification of objects which is more faithful to the principle of state encapsulation. CafeOBJ does not make any specific choice on the kind of object-orientation since it regards the object-orientation as a derived feature rather than a primary paradigm of the language. This gives the user the freedom to chose the most suitable form of object-orientation for their applications. However, the current CafeOBJ methodologies strongly emphasize the treatment of concurrent objects within behavioural specification [44] (see Sections 16 and 17).

Powerful module system

The principles of the CafeOBJ module system are inherited from OBJ which builds on ideas first realized in the language Clear [4, 5]. CafeOBJ has several kinds of imports, parameterized programming (also allowing integration of CafeOBJ specifications with executable code in a lower level language), views, and module expressions. However, the theory supporting the CafeOBJ module system represents an updating of the original Clear/OBJ concepts to the more sophisticated situation of multi-paradigm systems involving theory morphisms across institution embeddings [14], and the concrete design of the language revise the traditional OBJ view on importation modes and parameters.

Powerful type system

CafeOBJ has a type system that allows subtypes based on *order sorted algebra* (abbreviated **OSA**) [38, 30]. This provides a mathematically rigorous form of runtime type checking and error handling, giving CafeOBJ a syntactic flexibility comparable to that of untyped languages, while preserving all the advantages of strong typing. The order sorted feature of CafeOBJ not only greatly increases expressivity, but it might also provide a rigorous framework for multiple data representations and automatic coercions among them [30].

Since at this moment there are many order sortedness formalisms, many of them very little different from others, and each of them having its own technical advantages/disadvantages and being most appropriate for a certain class of applications, we decided to keep the concrete order sortedness formalism open at least at the level of the language definition. Instead we formulate some basic simple conditions which any concrete CafeOBJ order sortedness should obey. These conditions come close to Meseguer's OSA^R [48] which is a revised version of other versions of order sortedness existing in the literature, most notably Goguen's OSA [30].

Finally, CafeOBJ does not directly do partial operations but rather handles them indirectly by using error sorts and a sort membership predicate in the style of *membership equational logic* (abbreviated **MEL**) [48].

Logical Semantics

As with other languages from the OBJ family, CafeOBJ has logical semantics based on state-of-art developments in algebraic specification, the basic semantic document being [15]. The CafeOBJ Report can be regarded as unfolding all the details of the Logical Semantics in a rather concrete way.

The principles of the Logical Semantics can be formulated as:

(P1) there is an underlying logic[2] in which all basic constructs and features of the language can be rigorously explained,[3]

(P2) provide an integrated, cohesive, and unitary approach to the semantics of
programming/specification in the small and in the large, and

(P3) develop all ingredients (concepts, results, etc.) at the highest appropriate level of abstraction.

In order to achieve these principles we made extensive use of the powerful modern semantic tools made available by research in algebraic specification over the past decade, such as institutions and category theory. Institutions make it perfect for developing the semantics of sophisticated systems implementing a multitude of mutually interacting paradigms in a simple, clean, and compact manner. Modern systems, including CafeOBJ, cannot escape a certain degree of complexity and sophistication, however institutions (and more generally, categorical methods) greatly help in retaining a basic simplicity at least at the level of semantics. Moreover, our abstract logical approach permits future extensions of CafeOBJ with other paradigms provided they are rigorously based on logic and they interact well with the existing paradigms; such extensions will still lie within the present semantics.

---

[2]Here "logic" should be understood in the modern relativistic sense of "institution" which provides a mathematical definition for a logic (see [29]) rather than in the traditional sense.

[3]This intimate relationship between the language and its underlying logic is called "logical programming" by Goguen and Meseguer in [36].

The CafeOBJ Cube

The following table shows the correspondence between specification/programming paradigms and logics as they appear in the actual version of CafeOBJ, also pointing to some basic references.

ABBR.	LOGIC	SPEC/PGM PARADIGM	BASIC REF.
MSA	many sorted algebra	algebraic specification	[28]
OSA	order sorted algebra	algebraic specification with subtypes	[28, 38, 30]
HSA	hidden sorted algebra	behavioural concurrent specification	[31]
HOSA	hidden order sorted algebra	behavioural concurrent specification with subtypes	[31, 3]
RWL	rewriting logic	rewriting logic specification	[49]
OSRWL	order sorted rewriting logic	rewriting logic specification with subtypes	
HSRWL	hidden sorted rewriting logic	behavioural rewriting logic specification	[13, 10]
HOSRWL	hidden order sorted rewriting logic	behavioural rewriting logic specification with subtypes	

There are some enrichment/embedding relations between these logics, which correspond to institution embeddings [15, 14] (i.e., a strong form of institution morphisms of [29, 16]), and which are shown by the following CafeOBJ cube[4] (the orientation of arrows correspond to moving from "less complex" to "more complex" logics).

---

[4]The CafeOBJ cube is also the origin of the CafeOBJ logo.

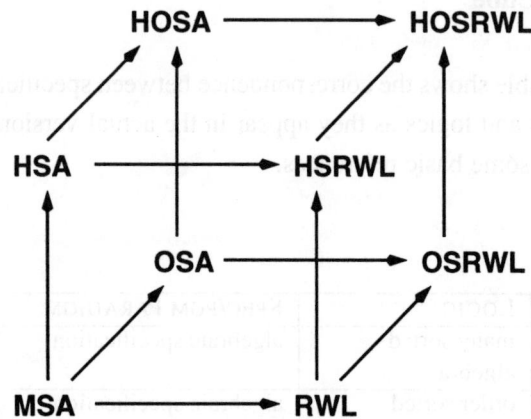

More rigorously, when dealing with pre-defined data types (required for any system having reasonable library support), the semantics must involve *constraint logics* [8, 11]. But since this issue is somehow secondary to our approach, and also because constraint logics can be easily internalized to any of the institutions constituting the CafeOBJ cube (see [11]), we feel that for the purpose of the presentation it is not necessary to add another dimension to the CafeOBJ cube.

HOSRWL embeds all other institutions (and therefore all main features of the language), hence it can be regarded as the institution underlying CafeOBJ. However, it is important to consider the CafeOBJ cube in its entirety rather than HOSRWL alone. In a sense, HOSRWL represents the flattening of the cube, and some subtle information on the relationship between the component features is lost in this flattening.[5]

Short Guide to Background Literature

Category theory plays a crucial rôle for the definition of CafeOBJ. Although this Report tries to avoid it as much as possible and be self-contained, a deeper

---

[5]One simple example is given by imports of MSA modules by RWL modules, their denotations should map RWL models to algebras by getting rid off the transitions. This process directly uses the embedding of MSA into RWL and cannot be explained within HOSRWL alone.

understanding of the design of CafeOBJ requires a good command of category theory. We recommend Mac Lane's categorical "bible" [45] for the readers of some mathematical strength, and any of the recent category theory for computer scientists textbooks (such as [1] or [55]) for ordinary readers.

Next comes the theory of institutions for which we recommend the seminal paper of Goguen and Burstall [29] as basic reading, and [14] for the new advanced concepts and results on institutions supporting the CafeOBJ design. A brief summary of the institution theory underlying the CafeOBJ definition can be found in this Report in the corresponding Appendix.

Finally, for a deeper understanding of the main paradigms of CafeOBJ we recommend [38, 30, 48] for OSA, [33] for HSA, [49] for RWL, and [10] for HSRWL. The basic reading for OBJ is [40]. As general algebraic specification reading we strongly recommend the coming Goguen's book [28] which contains both the basic stuff and all modern developments in this area, all of them presented in a very elegant and cohesive way.

I Basic Specifications

By **basic specification** we mean specifications not containing any structuring constructs (i.e., constructs related to the module system). In this part by "specification" we mean only basic specifications. The basic specifications can be classified into several kinds corresponding to the underlying paradigm (or institution, accordingly to the Logical Semantics) as follows:

- equational specifications,
- rewrite specifications,
- (simple) behavioural specifications, and
- behavioural rewriting specifications.

Some of the CafeOBJ constructs are specific only to some of these kinds of specifications, also the semantics of each construct is defined within a clearly established framework (i.e., institution; see the Institutions Appendix) corresponding to one of these kinds of specifications.

For each kind of specification we define the following concepts:

- a class $\mathbb{S}ign$ of signatures,
- for each signature Σ a class $\text{MOD}(\Sigma)$ of models,
- for each signature Σ a set $Sen(\Sigma)$ of sentences, and
- for each signature Σ a *satisfaction relation* $\models_\Sigma$ between Σ-models and Σ-sentences.

Specifications *SP* are practically finite collections of sentences for a given signature *sign(SP)*.

1

The Denotation of a Specification

Denotations of specifications represent the class of models (i.e., possible implementations[1]) of the respective specifications. They represent the semantics, or the meaning, of the specification. In fact, this is one of the characteristic features of algebraic specification: specifications are just formal descriptions of a certain class of models which exist only ideally as abstract mathematical objects. Therefore, understanding the denotations of CafeOBJ specifications can be considered to be one of the main goals of writing this Report.

In this part we deal only with denotations of *basic* specifications. CafeOBJ basic specifications can be classified into those with loose denotation, and those with tight denotation. The **loose denotation** of a specification consists of the class $\text{MOD}(SP)$ of all models of $sign(SP)$ which satisfy all sentences in SP. The **tight denotation** consists only of the **initial model** 0_{SP} in $\text{MOD}(SP)$, i.e., for each other model $M \in \text{MOD}(SP)$ of SP there exists a unique model morphism $0_{SP} \to M$. Initial models are unique up to isomorphisms, so essentially there is only one initial model. Therefore, in this Report, by "the" initial model we often mean the whole (isomorphism) class of initial models. CafeOBJ supports the distinction between loose and tight denotations by special keywords, mod! for tight semantics, and mod* for loose semantics.

The notation for the denotation of a specification SP is $[\![SP]\!]$. To resume

$$[\![SP]\!] = \begin{cases} 0_{SP} & \text{if tight (initial) semantics} \\ \text{MOD}(SP) & \text{if loose semantics.} \end{cases}$$

1   Signatures

Signatures are formed by a set of sorts and operations and predicates on this set of sorts.

Example 1 The following is a specification of the data type of natural numbers à la Peano.

```
mod! BARE-NAT {
```

---

[1] It is often very convenient to think of models as (possible) implementations of specifications, and of model morphisms as refinements between such implementations.

```
  [ NzNat Zero < Nat ]

  op 0 : -> Zero
  op s_ : Nat -> NzNat
}
```

The signature of the following specification describes a "history sensitive storage" object which stores natural numbers.

```
mod* HSS-BNAT {
  protecting(BARE-NAT)

  *[ Hss ]*

  bop put : Nat Hss -> Hss
  bop rest_ : Hss -> Hss
  bop get_ : Hss -> Nat
}
```

□

The second line of this module is an import declaration meaning that the specification BARE-NAT is included in HSS-BNAT. The other notations (for sorts and operations) are explained in the paragraphs below and formally presented in the syntax table Appendix.

1.1 Sorts

Sorts are names for entities of the same type, and they constitute the basis for the CafeOBJ type system. Semantically, sorts denote the set of entities of that type (sort); see Section 2.1. In CafeOBJ there are two kinds of sorts:

- ordinary (or *visible*) sorts (denoted by [ ]), used for specifying data types, and
- hidden sorts (denoted by * [ ] *), used (only) in behavioural specifications for specifying states of objects or abstract machines.

Their use is complementary, and they are always disjoint, in other words a sort in either ordinary or hidden, it cannot be both.

Example 2 In Example 1, Zero, NzNat, Nat are visible sorts, and Hss is hidden sort. □

CafeOBJ supports subtyping via the **subsort** construct which specifies an inclusion between the set of entities of a smaller type and the set of entities of a larger type. Subsorting is a partial order, i.e.,

1. $s \leq s$,
2. $s_1 \leq s_2$ and $s_2 \leq s_1$ implies $s_1 = s_2$, and
3. $s_1 \leq s_2$ and $s_2 \leq s_3$ implies $s_1 \leq s_3$.

for each sorts s, s_1, s_2, s_3, where "$\leq$" means "< or =", and ordinary and hidden sorts are always *not* related by the subsorting relation. A **connected component** of the subsorting relation is an equivalence class of the equivalence relation generated by this partial order.

Example 3 In Example 1, Zero and NzNat are declared as sub-sorts of Nat. The former contains the zero element, while the second contains all elements which are not zero. □

CafeOBJ supports error handling via **error sorts** which are denoted by ?*s* for some sort name *s*. For each connected component of the partial order of sorts there exists one error sort which is the upper bound for all sorts in this connected component. Error sorts in CafeOBJ should be regarded as playing the same rôle as *kinds* in MEL [48].

Example 4 ?Nat is the only error sort of BARE-NAT corresponding to its only sort connected component. The full partial order of the sorts of BARE-NAT is therefore as follows:

```
Zero NzNat < Nat < ?Nat
```

□

1.2 Operations and Predicates

Given a partially ordered set of sorts $(S, \leq)$, an **operation** σ on S is denoted as $\sigma\colon w \to s$ where $w \in S^*$ is its **arity** and $s \in S$ is its **sort** (sometimes called **co-arity**). The string ws is called the **rank** of the operation. **Constants** are operations whose arity is empty, i.e., $\sigma\colon [] \to s$.

Example 5 In Example 1, the arity of put is Nat Hss, its sort is Hss, and its rank is Nat Hss Hss. 0 is an example of constant of sort Zero, hence of sort Nat too. □

Operations can be **overloaded**, that is two different operations can be given the same name but only subject to the following condition:

> If $\sigma\colon w \to s$ and $\sigma\colon w' \to s'$ with w and w' of the same length, then we have that s is in the same connected component with s' if and only if w is (component-wise) in the same connected component with w'.

In OSA^{R} signatures obeying this condition are called *sensible* [48].

Each CafeOBJ specification inherits a built-in data type of Booleans, denoted Bool. Operations of sort Bool are called **predicates**.

In a signature Σ, the set of all operations of rank ws is denoted Σ_{ws}. The whole signature can be thus regarded as a family indexed by ranks, i.e., $\{\Sigma_{ws}\}_{w \in S^*, s \in S}$.

In behavioural specifications, some operations are declared as **behavioural operations**, and they are subject to the following syntactic condition:

> Behavioural operations have *exactly one* hidden sort in their arity.

Following the object-oriented terminology, behavioural operations are called **attributes** iff their sort is visible and **methods** iff their sort is hidden. The above condition then says that methods and attributes act on (states of) single objects. Behavioural operations are supported by the special CafeOBJ keyword bop.

Example 6 In Example 1, `put`, `rest`, and `get` are all behavioural operations, the first two are "methods", while the last is "attribute". □

CafeOBJ does not directly support **partial operations**, but rather handles them indirectly via error sorts and a sort membership predicate (see Section 4.1).

2 Models

Models lie at the heart of the semantics of algebraic specification since specifications describe models and, on the other hand, models can be regarded as implementations of specifications. Refinements between implementations are modeled by **model morphisms**.

In this section we present several kinds of models corresponding to the three major paradigms, equational, rewriting, and behavioral specification. Models of some specification paradigm can be reduced to models of a simpler paradigm; this reduction corresponds to the opposite of the arrows in the CafeOBJ cube.

2.1 Algebras

Given a signature $(S, \leq, \Sigma)$, an **algebra** A interprets

(A1) each sort $s \in S$ as a set A_s,

(A2) each subsort relation $s < s'$ as an inclusion $A_s \subseteq A_{s'}$, and

(A3) each operation $\sigma \in \Sigma_{s_1 \ldots s_n s}$ as a function $A_\sigma \colon A_{s_1} \times \ldots \times A_{s_n} \to A_s$.

An **algebra morphism** $f \colon A \to B$ is an S-indexed family of functions between the carriers of A and B, $\{h_s \colon A_s \to B_s\}_{s \in S}$, such that

(M1) if $s < s'$, then $h_{s'}(x) = h_s(x)$ for all $x \in A_s$, and

(M2) $h_s(A_\sigma(a_1, \ldots, a_n)) = B_\sigma(h_{s_1}(a_1), \ldots, h_{s_n}(a_n))$ for all $\sigma \in \Sigma_{s_1 \ldots s_n s}$, and $a_i \in A_{s_i}$ for $i \in [n]$.

A **congruence** $\equiv$ on an algebra A is an S-sorted equivalence on A (i.e., an equivalence $\equiv_s$ on A_s for each sort $s \in S$) such that

(C) if $a_i \equiv_{s_i} a_i'$ for $i \in [n]$ then $A_\sigma(a_1, \ldots, a_n) \equiv_s A_\sigma(a_1', \ldots, a_n')$ for all $\sigma \in \Sigma_{s_1 \ldots s_n s}$ and for all ranks $s_1 \ldots s_n s$.

Example 7 For the specification BARE-NAT of Example 1 we can have several different models. The most obvious one is **Nat** which interprets

- **Nat**$_{\text{Nat}}$ as ω the set of natural numbers, **Nat**$_{\text{Zero}}$ as $\{0\}$, and **Nat**$_{\text{NzNat}}$ as $\omega - \{0\}$ the set of strictly positive natural numbers,
- **Nat**$_0$ as 0, and **Nat**$_s(n) = n + 1$ for all $n \in \omega$.

The denotation $[\![\text{BARE-NAT}]\!]$ consists only of **Nat** which is the initial model (this concept will be fully explained in Section 4.1). A more eccentric model is **Strg** interpreting

- **Strg**$_{\text{Nat}}$ as $\{a, b\}^*$ the set of strings formed with a and b, **Strg**$_{\text{Zero}}$ as the set containing only the empty string $\{[]\}$, and **Strg**$_{\text{NzNat}}$ as the set of non-empty strings $\{a, b\}^+$,
- **Strg**$_0$ as $[]$, and **Strg**$_s(str) = a.str$, i.e., concatenation with a, for all strings str.

Notice that there is an (unique) algebra morphism h: **Nat** $\to$ **Strg** mapping each $n \in \omega$ to the string a^n formed of n of a's.

There are also several congruences on **Strg**, such as

1. $str \equiv^a str'$ iff str and str' contain the same number of a's,
2. $str \equiv^b str'$ iff str and str' contain the same number of b's

□

2.2 Rewrite models

Rewrite models give a nice mathematical formalization for the notion of concurrent distributed system [49]. The carriers of a rewrite model contain two different kinds of entities:

- *states*, modeled as categorical objects, and
- *transitions* (between states), modeled as categorical arrows.

By "categorical objects and arrows" we mean objects and arrows as precise mathematical entities in the sense of category theory [45]. As already noticed, categories are one of the fundamental mathematical structures for the semantics of CafeOBJ and they are used in several different ways. Here, **categories** are used for modeling systems configurations in terms of states (as objects) and transitions (as arrows), which is the fundamental idea of the semantics of rewriting logic [49].

This means that each transition has a state as source and another as target, that transitions compose associatively (see the diagram below), and that each state has a trivial transition which is identity for the composition of transitions.

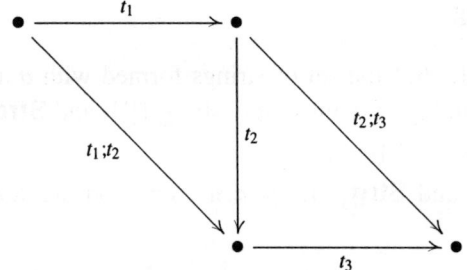

Mappings h between carriers of rewrite models map states st to states $h(st)$ and transitions $st \xrightarrow{t} st'$ to transitions $h(st) \xrightarrow{h(t)} h(st')$ such that identity transitions and the composition of transitions are preserved, i.e., $h(t;t') = h(t);h(t')$ for all transitions t, t' which compose (the target of t being the same with the source of t'). This is exactly the categorical notion of **functor** (see [45]).

Given a signature $(S, \leq, \Sigma)$, a **rewrite model** A interprets

(R1) each sort $s \in S$ as a *category* A_s,

(R2) each subsort relation $s < s'$ as an inclusion of categories $A_s \subseteq A_{s'}$, and

(R3) each operation $\sigma \in \Sigma_{s_1 \ldots s_n s}$ as a functor $A_\sigma : A_{s_1} \times \ldots \times A_{s_n} \to A_s$.

Morphisms of rewrite models are S-indexed families of *functors* satisfying conditions M1 and M2. Rewrite model congruences are S-sorted **category equivalences** (i.e., equivalences on both states and transitions) satisfying the congruence condition (C).

Rewrite models are "algebras over categories" rather than over sets as in the ordinary equational logic case. Consequently, functors play for rewrite models the same role functions play for ordinary algebras. Each rewrite model can be reduced to an algebra by forgetting the transitions together with all adjacent structures, i.e., by regarding categories as sets (of objects) and functors as functions (between sets of objects).

Example 8 Consider the following rewrite model **SStrg** for the specification BARE-NAT of Example 1 for which

- the states are strings, i.e., $|\mathbf{SStrg}_{sort}| = \mathbf{Strg}_{sort}$ for all sorts $sort \in \{\texttt{Zero}, \texttt{NzNat}, \texttt{Nat}\}$,

- the transitions between two strings $strg$ and $strg'$ intuitively erase some parts of the former string for getting the latter, and are formally defined (for each sort $sort$) by

$$\mathbf{SStrg}_{sort}(strg, strg') = \{f : [\#strg'] \to [\#strg] \mid f \text{ strictly increasing and } strg'_i = strg_{f(i)} \text{ for all } i \in [\#strg']\}$$

where by # we denote the length of a string and by $strg_i$ we denote the i-th element of the string $strg$,

- $\mathbf{SStrg}_s$ extends $\mathbf{Strg}_s$ to a functor by

$$\mathbf{SStrg}_s([\#strg'] \xrightarrow{f} [\#strg]) = [\#a.strg'] \xrightarrow{f'} [\#a.strg], \text{ where } f'(1) = 1$$
and $f'(n) = f(n-1) + 1$ for $2 \le n$.

Notice that the rewrite model **SStrg** is reduced exactly to the algebra **Strg** by forgetting the transitions and all adjacent structures. $\square$

2.3 Hidden models

Given a signature $(S, \le, \Sigma)$ with a subset $H \subseteq S$ of hidden sorts, a **hidden model** M (which can be either an algebra or rewrite model) interprets the (set of) visible sorts V and the (set of) operations on the visible sorts ψ as a *fixed* model D (called the **model of data**) and the hidden part of the signature arbitrarily, that is $M\restriction_{V,\psi} = D$ by using the notation of reduct (see Section 6). A **morphism of hidden models** $h: M \to M'$ preserves the data, i.e., $h\restriction_{V,\psi} = 1_D$.

In most cases the model of data is the standard (initial) model of a specification. Notable exception cases are some pre-defined (built-in) data types (see Section 11), such as real numbers. A **hidden relation** R on a hidden model M is just a many sorted binary relation $\{R_s\}_{s \in S}$ on M which is the identity on the visible sorts.

Example 9 Consider the specification of the "history sensitive storage" of Example 1. We fix the model of data to be **Nat** of Example 7. From the possible models of HSS-BNAT, we restrict our attention to two of them.

The first one, **List**, interprets

- **List**$_{\text{Hss}}$ to be the set of all lists of natural numbers, such that
- **List**$_{\text{put}}$ is concatenating an element to the beginning of a list, **List**$_{\text{rest}}$ is taking the tail of a non-empty list and does nothing on the empty list, and **List**$_{\text{get}}$ is taking the head of a non-empty lists and returns 0 for the empty list.

The second model is an "array with pointer" model **Arr** such that:

- **Arr**$_{\text{Hss}} = \omega^\omega \times \omega$, i.e., arrays are pairs of functions from natural number indices to natural numbers and indices (for pointers),
- **Arr**$_{\text{put}}(a, \langle f, n \rangle) = \langle f', n+1 \rangle$, where $f'(x) = f(x)$ for $x \neq n$ and $f'(n) = a$, **Arr**$_{\text{get}}(\langle f, n \rangle) = f(n-1)$, and **Arr**$_{\text{rest}}(\langle f, n \rangle) = \langle f, n-1 \rangle$ if $n \geq 1$ and $\langle f, 0 \rangle$ otherwise.

Notice that there is a "quotient" hidden model morphism $q \colon$ **Arr** $\to$ **List** given by

$$q(\langle f, n \rangle) = f(0).f(1).\ldots.f(n-1)$$

$\square$

Each hidden model is in fact an ordinary model which interprets the data part in a fixed way. This also means that each hidden rewrite model can be reduced to a hidden algebra.

3 Sentences

3.1 Terms

Given a signature $(S, \leq, \Sigma)$, the basic constituents of sentences for this signature are the **terms** formed with operators from Σ and variables for the sorts S. The variables of a specification over the signature $(S, \leq, \Sigma)$ form an S-sorted set; each variable is declared to belong to a specific sort in S.

Given an S-sorted set $X = \{X_s\}_{s \in S}$ of variables, a Σ-**term** is defined recursively as follows:

- each constant $\sigma \in \Sigma_s$ is a Σ-term of sort s,

- each variable $x \in X_s$ is a Σ-term of sort s,

- t is a term of sort s' if t is a term of sort s and $s < s'$, and

- $\sigma(t_1, \ldots, t_n)$ is a term of sort s for each operation $\sigma \in \Sigma_{s_1 \ldots s_n s}$ and terms t_i of sort s_i for $i \in [n]$.

The S-sorted set of Σ-terms is denoted as $T_\Sigma(X)$. A Σ-term is **ground** if it does not contain any variables.

The Σ-terms (modulo renaming of the variables) can be regarded as **derived operations** by defining arity and rank for terms t by the following procedure:

- consider the set $var(t)$ of all variables occuring within t,

- transform $var(t)$ into a string by fixing an arbitrary order on this set, and

- finally, replace the variables in the string previously obtained by their sorts.

The rank of a term is the concatenation of its arity and its sort. A derived operation (i.e., term) is **behavioural** iff it does not contain any (primitive) operation on[2] hidden sorts which is not behavioural.

So, given a signature $(S, \leq, \Sigma)$, we may consider the **derived signature** $(S, \leq, \Sigma^*)$ where Σ_{ws}^* is the set of all Σ-terms modulo renaming of variables

---

[2]Notice the difference between "on" and "of" here: "on" refers to the *arity* of the operation, while "of" refers to the *sort* of the operation.

with arity w and sort s. The following principle governs the relationship between derived and primitive operators:

> We identify the Σ^*-terms with the Σ-terms by unfolding the derived operations.

3.2 Equations

Conceptually, equations are equational specification sentences. However they also appear in rewriting and behavioural specifications since they both extend the equational specification paradigm.

Given a signature $(S, \leq, \Sigma)$, then an **equation** is defined as

$$(\forall X)\, t = t' \ \textbf{if}\ C$$

where X is an S-sorted set of variables, t, t' are Σ-terms in $T_\Sigma(X)$, and C is a Σ-term of sort Bool. $(\forall X)$ is called the **quantifier** of the equation, $t = t'$ is the **body** of the equation, and C is the **condition** of the equation. When the condition is missing the equation is called **unconditional**, and is written as

$$(\forall X)\, t = t'$$

Due to purely operational (semantics) reasons, CafeOBJ equations are subject to the following syntactic restriction:

$$var(t') \subseteq var(t) \ \text{ and } \ var(C) \subseteq var(t)$$

where $var(\_)$ is the set of variables occurring in the term. The quantifier is not explicitly declared in CafeOBJ equations, but rather assumed by convention to be the set of all variables occurring in the equation (i.e., $var(t)$ with the notations used above).

Examples of equations appear beginning with Example 11.

Operation attributes

A special case of equations is given by the **operation attributes**. Although at the denotational level the operation attributes can be treated as ordinary equations, they have special operational semantics (see Section 12).

CafeOBJ supports 4 basic binary operation attributes and their combinations:

- *associativity*, $(\forall\{x,y,z\})\ \sigma(x,\sigma(y,z)) = \sigma(\sigma(x,y),z)$,

- *commutativity*, $(\forall\{x,y\})\ \sigma(x,y) = \sigma(y,x)$,

- *identity*, $(\forall\{x\})\ \sigma(x,e) = \sigma(e,x) = x$ for some constant e, and

- *idempotence*, $(\forall\{x\})\ \sigma(x,x) = x$.

where σ is a binary operation.

Example 10 Monoids can be specified by using the associativity and identity attributes:

```
mod* MON {
   [ Elt ]

   op nil : -> Elt
   op _;_ : Elt Elt -> Elt {assoc id: nil}
}
```

We can further specify the commutative monoids by adding the commutativity attribute:

```
mod* CMON {
   protecting(MON)

   op _;_ : Elt Elt -> Elt {comm}
}
```

□

3.3 Transitions

Transitions are sentences specific only to rewrite specifications, and syntactically they can be regarded as oriented equations. In the literature [49, 47] they are better known under the name of *rules*. However, their semantics is different from that of equations. As in the case of equations, a **transition** is given by

$$(\forall X)\, t \;=>\; t' \;\textbf{if}\; C$$

In the absence of the condition we get unconditional transitions

$$(\forall X)\, t \;=>\; t'$$

Transitions are subject to the same syntactic restrictions and conventions with the equations.

Examples of transitions appear beginning with Example 13.

3.4 Behavioural sentences

Behavioural sentences can be either equations or transitions. Syntactically they are the same as ordinary sentences, but they are distinguished from the ordinary ones by the special CafeOBJ keywords with "b" as prefix.[3] As with ordinary sentences, conditions of behavioural conditional sentences are just Bool-sorted terms.

The real difference between ordinary and behavioural sentences is that the latter specify *behavioural* properties, so their satisfaction by models is different (i.e., behavioural rather than strict; see Section 4.3). Examples of behavioural sentences appear beginning with Example 14.

4   Satisfaction

The satisfaction relation between models and sentences goes to the heart of semantics of algebraic specification. This asserts if a certain implementation has a certain property (written as a syntactic formula, or sentence, in CafeOBJ). Each of the basic paradigms of CafeOBJ has its own notion of satisfaction.

---

[3]For example, "beq" stands for behavioural equations, "btrans" for behavioural transitions, etc.

Valuations and term interpretation

Valuations assign values to variables, in other words they represent instantiations of the variables with values from a given model. Given a signature $(S, \leq, \Sigma)$, a model M for it, and an S-sorted set X of variables, a valuation $\theta \colon X \to M$ consists of an S-sorted family of maps $\{\theta_s \colon X_s \to M_s\}_{s \in S}$. In the case of rewrite models it is convenient to distinguish between

- *state valuations* interpreting variables as states, i.e., $\theta_s \colon X_s \to |M_s|$ for each $s \in S$, and

- *transition valuations* interpreting variables as transitions, i.e., $\theta_s \colon X_s \to M_s$ for each $s \in S$.

Each Σ-term t over the variables X can be interpreted as a value $\theta(t)$ in the model M for each valuation $\theta \colon X \to M$ in the following inductive manner:

- M_σ if t is a constant σ,

- $\theta(x)$ if t is a variable x,

- $M_\sigma(\theta(t_1), \ldots, \theta(t_n))$ if t is of the form $\sigma(t_1, \ldots, t_n)$ for some $\sigma \in \Sigma_{s_1 \ldots s_n s}$ and terms t_i of sort s_i.

Notice that in the case of rewrite models a term is interpreted as a state or as a transition depending on whether the valuation used is a state valuation or, a transition valuation, respectively.

In this way, given a model M, each (not yet evaluated) term t can be interpreted as a function in the equational case, or as a functor in the RWL case, i.e., $M_t \colon M_{s_1} \times \ldots \times M_{s_n} \to M_s$ where $s_1 \ldots s_n$ is the arity of t and s is its sort.

4.1 Equational Specification

We fix a signature $(S, \leq, \Sigma)$. An equation $(\forall X)\, t = t'$ **if** C is **satisfied** by a Σ-algebra M, denoted as

$$M \models (\forall X)\, t = t' \text{ if } C$$

iff $\theta(t) = \theta(t')$ whenever $\theta(C) = M_{\text{true}}$ for all valuations $\theta \colon X \to M$.

Informally, an equation is satisfied by an algebra iff all possible ways to assign values to variables evaluate both sides of the equation as the same value, with proviso that the condition C is satisfied.

The initial algebra of an equational specification

Notice that given a signature $(S, \leq, \Sigma)$ and a set of variables X, the Σ-terms $T_{\Sigma}(X)$ can be organized as a Σ-algebra in the obvious way (by using the inductive definition of Σ-terms). When X is empty, $T_{\Sigma}(\emptyset)$ is denoted as T_{Σ}, and has the following *initiality* property (see [28, 38])

> for each algebra M there exists a unique Σ-morphism
> $T_{\Sigma} \to M$.

Given a set E of equations for $(S, \leq, \Sigma)$, then we construct the algebra $T_{\Sigma, E}$ as follows

- for each $s \in S$ let $(T_{\Sigma, E})_s$ be the set of equivalence classes of Σ-terms in T_{Σ} under the congruence $\equiv^E$ defined as $t \equiv^E t'$ iff $(\forall \emptyset)\, t = t'$ can be deduced from E using the proof system of equational specification (see Section 5), and

- each operation $\sigma \in \Sigma_{s_1...s_n s}$ is interpreted as
  $(T_{\Sigma, E})_{\sigma}(t_1/\equiv^E, \ldots, t_n/\equiv^E) = \sigma(t_1, \ldots, t_n)/\equiv^E$ for all $t_i \in (T_{\Sigma})_{s_i}$ (for $i \in [n]$) by using the property of $\equiv^E$ as congruence on T_{Σ}.

$T_{\Sigma, E}$ has the following *initiality* property (see [28, 38])

> for each algebra M *satisfying all equations in E*,
> there exists a unique Σ-morphism $T_{\Sigma, E} \to M$.

and it is the model giving the tight denotation of the equational specification E.

Example 11 Consider the following CafeOBJ specification of the data structure of strings of natural numbers:

```
mod! STRG-BNAT {
  protecting(BARE-NAT)

  [ Nat < Strg ]

  op nil : -> Strg
  op _._ : Strg Strg -> Strg {assoc}

  var S : Strg
  var N : Nat

  eq S . nil = S .
  eq nil . S = S .
}
```

The initial model **StrgNat** of STRG-BNAT is defined by

- **StrgNat$_{Strg}$** is ω^*, i.e., the set of (finite) strings of natural numbers, and
- **StrgNat$_{nil}$** is the empty string $[]$, and **StrgNat$_{\_\_}$** concatenates strings.

The initiality property for **StrgNat** follows by a routine proof by induction on the structure of strings (we leave this as exercise to the reader). Because of mod! which means a tight semantics declaration, **StrgNat** is the only model of STRG-BNAT. $\Box$

The built-in predicate _==_

CafeOBJ implements *semantic* equality (i.e., equality of model values) as a many-sorted binary relation

$$\_==\_ \ : s \ s \rightarrow \text{Bool}$$

for each sort $s \in S$.[4]

In the case of the initial model $T_{\Sigma,E}$ for an equational specification (Σ, E), the semantic equality is generally semi-decidable since

---

[4]In fact the condition on the arguments of _==_ can be relaxed to sorts belonging to the same connected component, the SRA implementation of CafeOBJ following this idea [51]. But for simplicity we restrict the presentation to the "same sort" approach.

$t == t'$ iff $(\forall \emptyset)$ $t = t'$ is a (proof-theoretic) consequence of the equations in E.

However, very frequently, when E is a *(ground) confluent and terminating* rewrite system (see Section 12), $\_==\_$ on $T_{\Sigma,E}$ is fully decidable. CafeOBJ evaluates $\_==\_$ by reducing both sides to normal forms by using the command reduce and then checking whether these are identical.

The predicate $\_==\_$ is frequently used for encoding classical conditions of equations (which are finite conjunctions of equalities between terms) as CafeOBJ conditions by translating the syntactic equality into the semantic equality predicate $\_==\_$ and by using the built-in Bool-operation and for expressing the conjunctions between equalities.

The sort membership predicate $\_:s$

In order to handle partial functions, CafeOBJ implements a (family of) **sort membership** predicates. For each error sort $?s$, and for each sort s' in the connected component of $?s$, there is a predicate

$$(\_:s') : ?s \rightarrow \text{Bool}$$

which gets evaluated to true if and only if its argument belongs to the sort s'. This means that the sort membership predicates might take part in assigning the (least) sort for terms, which means that the parsing process involves not only the data given by the signature declarations, but also equations from the specification. This extension is fully explained by the MEL framework [48].

The sort membership predicate can be used in conjunction with error sorts for handling partial functions; the following is a typical example:

Example 12 Consider the specification of path in a directed graph (each edge or path has a "source" and a "target").

```
mod* GRAPH {
  [ Node Edge ]

    ops (s_) (t_) : Edge -> Node
}
```

The path concatenation is a partial function with an equationally defined domain. In this specification the error sort contains all possible expressions obtained from applying the concatenation operation indiscriminately, but only
real path can "parse" as having sort Path by using the sort membership predicate.

```
mod! PATH (G :: GRAPH) {
  [ Edge < Path ]

  op _;_ : ?Path ?Path -> ?Path   {assoc}
  ops (s_) (t_) : Path -> Node

  var E : Edge
  var EP : ?Path

  ceq (E ; EP) : Path = true
          if (EP : Path) and (s EP) == (t E) .
  ceq s(E ; EP) = s(E)  if (E ; EP) : Path .
  ceq t(E ; EP) = t(EP) if (E ; EP) : Path .
}
```

Notice that the full PATH contains the sort membership predicates

```
  ops (_: Edge) (_: Path) : ?Path -> Bool
```

and the following equations

```
  var E : Edge
  var P : Path
  eq E : Edge = true .
  eq P : Path = true .
```

□

4.2 Rewrite Specification

We again fix a signature $(S, \leq, \Sigma)$. In rewrite specification, the satisfaction between rewrite models and sentences has two different aspects: the satisfaction
of equations, and the satisfaction of transitions.

The satisfaction of equations by rewrite models has the same definition as for equational specification, with the mention that we have to consider both state and transition valuations.

However the satisfaction of transitions is different since transitions are sentences about unidirectional changes rather than equality. A transition $(\forall X)\, l \Rightarrow r$ **if** C is satisfied by a model M,

$$M \models (\forall X)\, l \Rightarrow r \ \textbf{if}\ C$$

iff for each state valuation $\theta\colon X \to |M|$ there is a transition $\alpha_\theta\colon \theta(l) \to \theta(r)$ whenever $\theta(C) = M_{\texttt{true}}$ such that for all transition valuations $\varphi\colon X \to M$ we have $\alpha_\theta; \varphi(r) = \varphi(l); \alpha_{\theta'}$,

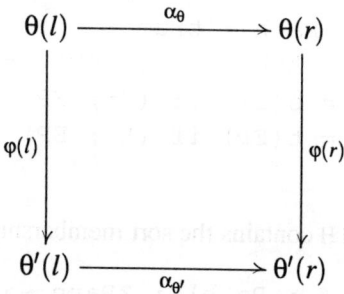

where $\varphi(x)\colon \theta(x) \to \theta'(x)$ for all variables $x \in X$. This commutativity property says that α is a **natural transformation** (see [49]) $M_l \Rightarrow M_r$, where M_l and M_r are the functors $M^X \to M$ determined by the terms l and r, and where by M^X we denote the valuations $X \to M$. Notice that the diagram above expresses this property for a transition valuation φ which has θ as source state valuation and θ' as target state valuation. This means that α can be regarded as a *global* transition (parameterized by the states) in a distributed system.

The initial model of a rewrite specification

As with equational specification, rewrite specification admits initial models as tight semantics. Given a signature $(S, \leq, \Sigma)$, let E be a set of equations and R be a set of Σ-transitions. Then let $T_{\Sigma, E, R}$ be the rewrite model such that

- $|(T_{\Sigma,E,R})_s| = (T_{\Sigma,E})_s$ for each sort $s \in S$, i.e., the states of the sort s are exactly equivalent classes of Σ-terms without variables under the equations E,

- $T_{\Sigma,E,R}(t/\equiv^E, t'/\equiv^E)$ is the set of all possible concurrent rewrites of $t/\equiv^E$ into $t'/\equiv^E$ by using the transitions in R modulo the equations E. This is the same as the set of all inference chains for deducing $(\forall \emptyset)\, t \Rightarrow t'$ from R and E by using the proof system of rewriting logic (see Section 5), and

- each operation $\sigma \in \Sigma_{s_1 \ldots s_n s}$ is interpreted

   - on states, as in the case of the algebra $T_{\Sigma,E}$, and
   - on transitions, in the following way: if $t_i/\equiv^E \xrightarrow{r_i} t'_i/\equiv^E$, for $i \in [n]$, are concurrent rewrites modulo E, then

     $(T_{\Sigma,E,R})_\sigma(r_1, \ldots, r_n)$ is the concurrent rewrite
     $\sigma(t_1, \ldots, t_n)/\equiv^E \to \sigma(t'_1, \ldots, t'_n)/\equiv^E$ modulo E which uses r_i, for $i \in [n]$.

Notice that the interpretations of the operations on transitions use terms over transitions as identifiers for the transitions in the model, where primary identifiers are the labels of the specification transitions, and complex identifiers are built as terms over these labels.

$T_{\Sigma,E,R}$ has the following *initiality* property (see [49]):

for each rewrite model M *satisfying all equations in E* and *all transitions in R*, there exists a unique Σ-morphism of rewrite models $T_{\Sigma,E,R} \to M$,

and this model gives the tight denotation of the rewrite specification (Σ, E, R).

Example 13 We can specify strings with unreliable concatenation (in the sense that an element which is concatenated to a string might be lost) by refining the strings Example 11

```
mod! USTRG-BNAT {
    protecting(STRG-BNAT)
```

```
    trans N:Nat . S:Strg => S .
}
```

by adding a transition sentence stating the "losing" property. The initial (rewrite) model **UStrgNat** of USTRG-BNAT refines the initial model **StrgNat** of STRG-BNAT by adding transitions between the strings of natural numbers such that

$$\textbf{UStrgNat}_{\textbf{Strg}}(strg1, strg2) = \{del : strg1 \rightarrow strg2 \mid del \text{ is a sequence}$$
of deletions from *strg1* which gets *strg2*}

for all strings *strg1* and *strg2*. For example, from 1.1.1 to 1 there are six possible sequences of deletions corresponding to all combinations of two positions out of the three possible positions in the string 1.1.1. Notice that these deletions are considered *sequentially* rather than in parallel.

The composition of sequences of deletions between strings is the obvious one, thus making **UStrgNat**$_{\textbf{Strg}}$ into a category.

UStrgNat$_{\_\_}$ concatenates (vertically, or in parallel) sequences of deletions. For example the concatenation $t_1.t_2$ of $t_1 : 1.2 \rightarrow 2$ with $t_2 : 3.1 \rightarrow 1$ is the transition $1.2.3.1 \rightarrow 2.1$ erasing the first 1 and the 3. $\square$

The built-in predicate _==>_

As with equality in equational specification, in rewrite specification there is a semantic counterpart of the syntactic transition symbol =>. CafeOBJ implements *semantic* transitions as a many-sorted binary predicate

$$\_==>\_ \quad : s \ s \rightarrow \texttt{Bool}$$

for each *visible* sort $s \in S$.[5] The semantic transition predicate is not defined on the hidden sorts, the main reason being to prevent an ad-hoc mixture of the RWL concurrency concepts (such as state and transition) with the behavioural ones.

The meaning of this transition predicate is "there exists at least one transition between the two argument states". This means that the predicate _==>_ collapses (possible) multiple transitions between states to only one transition.

---

[5]As with _==_, the arguments of _==>_ need not be of the same sort, their sorts can be in the same connected component.

This simplification gives CafeOBJ special proof strength, also avoiding the complication of Maude conditional transitions.[6]

The definition of $\_==>\_$, provided automatically by CafeOBJ in the case of rewrite specifications, encodes the reflexivity, congruence, and transitivity rules from the proof theory of rewriting logic, as well as all user defined transitions:

1. $x ==> x =$ true

   for all sorts $s \in S$ and all variables x of sort s,

2. $\sigma(x_1,\ldots,x_n) ==> \sigma(x'_1,\ldots,x'_n) =$ true
   if $(x_1 ==> x'_1)$ and ... and $(x_n ==> x'_n)$
   for all $\sigma \in \Sigma_{s_1\ldots s_n s}$ a non-behavioural operation and x_i, x'_i variables of sort s_i, for $i \in [n]$, and

4. $l ==> r =$ true **if** C

   for each transition $(\forall X)\, l => r$ **if** C.

CafeOBJ provides support for the transitivity of $\_==>\_$ via related predicates with strong operational flavour.[7] However, due to the very difficult computational nature of transitions (difficulty especially due to the lack of confluence), the implementation support for RWL transitions is bound to be rather limited. Notice also that the encodings of the user-defined transitions (4.) are used via substitutions generated by the matching mechanism of the CafeOBJ execution engine.

As in the case of conditional equations, $\_==>\_$ provides the support for encoding classical conditional transitions (in which conditions are finite conjunctions of [syntactic] transitions) into CafeOBJ conditional transitions by

---

[6]Dealing with multiple transitions would require the further complexity of *labeled* transitions as in Maude and would require heavy categorical machinery for defining conditional transition satisfaction.

[7]They are outside the scope of the present Report; also this support might vary between various implementations of CafeOBJ. For example, the SRA implementation [51], supports a built-in predicate

$\_=()=>\_$ $: s$ Nat $s \to$ Bool

encoding the possibility of n-steps transitions.

translating the syntactic transitions into predicates `_==>_` and by using the built-in `Bool`-operation `and` for expressing the conjunctions between transitions.

4.3 Behavioural Specification

Behavioural specification satisfaction (**behavioural satisfaction**, for short) establishes properties that do not hold in the usual strict sense, but which hold under **behavioural equivalence (equality)**.

Behavioural satisfaction is therefore defined on top of a given satisfaction relation. Given a signature $(S, \leq, \Sigma)$ with a subset $H \subseteq S$ of hidden sorts and a model D of data, let M be a model (either algebra or rewrite model) and ρ be a sentence (either equation or transition) not containing any operation on hidden sorts which is not declared behavioural. Then

$$M \models\!\!\!| \ \rho \quad \text{iff} \quad \overline{M} \models \rho$$

where $\models\!\!\!|$ denotes behavioural satisfaction, $\models$ denotes the strict (ordinary) satisfaction of sentences, and $\overline{M}$ is the behavioural image of M. The **behavioural image** of a model forgets (i.e., ignores) the interpretation of operations on hidden sorts which are not declared behavioural, and

- identifies the behavioural equivalent elements (states),
- identifies the behavioural equivalent transitions, and
- introduces (new) behavioural transitions.

These concepts are explained in the following paragraphs.

Behavioural equivalence $\sim$ (the corresponding semantic relation in CafeOBJ is denoted as `=b=`) between entities of hidden sorts is established as equality under all possible visible observations (experiments), i.e.,

$$a \sim a' \quad \text{iff} \quad M_c(a) = M_c(a')$$

for all **behavioural Σ-contexts** c. Σ-**Contexts** are Σ-terms with only one occurence of only one variable; **behavioural Σ-contexts** are Σ-contexts of visible sort without any operation on hidden sorts which is not declared behavioural on top of the variable. Behavioural contexts can be regarded as strings of "method" applications followed by an "attribute" evaluation.

A basic result in behavioural specification (see [3, 33]) showing that

> Behavioural equivalence is the largest hidden congruence[8]
> (with respect to behavioural operations).

provides foundation for the proof theory of behavioural specification. In the case of algebras, the behavioural image is exactly the quotient under behavioural equivalence. However, behavioural image of rewrite models might introduce new transitions.

Behavioural satisfaction can be extended to sentences containing operations on hidden sorts which are not declared behavioural in two different but still equivalent ways. The first one is by noticing that if all non-behavioural operations on hidden sorts are **behaviourally coherent** (i.e., they preserve behavioural equivalence), then the behavioural image is canonically a model for the full signature by interpreting canonically the non-behavioural operations on the congruence classes under behavioural equivalence. Then the above definition of behavioural satisfaction can be used for sentences containing non-behavioural operations on hidden sorts.

The second way is slightly more general since it does not rely on the behavioural equivalence being a congruence with respect to non-behavioural operations on hidden sorts:

$$M \models (\forall X)\, t = t' \text{ iff } \theta(t) \sim \theta(t')$$

for all valuations $\theta\colon X \to M$. Although the behavioural coherence of non-behavioral operations on hidden sorts is not necessary for defining behavioural satisfaction, it is still a necessary condition for the soundness of the ordinary equational inference system for behavioural equivalence.

Alternatively, a syntactic equivalent version of behavioural satisfaction is given by

$$M \models (\forall X)\, t = t' \text{ iff } M \models (\forall X)\, c(t) = c(t')$$

---

[8]A hidden relation which is a congruence.

and

$$M \models (\forall X)\, t \implies t' \text{ iff } M \models (\forall X)\, c(t) \implies c(t')$$

for all behavioural contexts c with arity matching the sort of t and t'. This definition of behavioural satisfaction extends easily to the conditional case just as in the case of ordinary satisfaction.

Example 14 Consider the following behavioural specification of a "history sensitive storage" of natural numbers extending the signature of Example 1 with equations:

```
mod* HSS-BNAT {
  protecting(BARE-NAT)

  *[ Hss ]*

  bop put : Nat Hss -> Hss
  bop rest_ : Hss -> Hss
  bop get_ : Hss -> Nat

  var S : Hss
  var E : Nat

  eq get put(E, S) = E .
  beq rest put(E, S) = S .
}
```

From the possible models of HSS-BNAT, let's pay attention to the two candidates of Example 9.

       List $\models$ HSS-BNAT therefore **List** $\models$ HSS-BNAT

We also have that

       Arr $\models$ HSS-BNAT

but notice that this model does satisfy the last equation *only* behaviourally, it does not satisfy it in the strict equational sense. Moreover, **List** is the *behavioural image* of **Arr** under the quotient morphism q.

We can specify a second order function standing for arbitrary long sequences of applications of rest[9] by "currying" it.

```
mod HSS-BNAT-PROOF {
  protecting(HSS-BNAT)

  bop rest* : Hss Nat -> Hss

  var S : Hss
  vars N : Nat

  eq [ p1 ] : rest*(S, s(N)) = rest*(rest S, N) .
  eq [ p2 ] : rest*(S, 0) = S .
}
```

Notice that these two equations express *real equalities* between elements of the hidden sort Hss rather than behavioural equalities because rest* represents an abbreviation for sequences of applications of rest rather than a new "method". □

The built-in hidden equivalence _=b=_

The CafeOBJ semantic relation denoting behavioural equivalence is written as =b= and is defined only on the hidden sorts:

$$\_=b=\_ \quad : h \ \ h \to \ \text{Bool}$$

Its use and implementation is strongly related to the execution command breduce (see Section 12.2), and is similar to _==_ in the sense that

$$t =b= t' = \textbf{true}$$

iff both t and t' reduce to the same normal form by using the command breduce. Notice that _=b=_ is not even semi-decidable.

---

[9]This will later be essentially used in some proofs about HSS-BNAT.

The built-in hidden equivalence _=*=_

CafeOBJ implements a default hidden relation which in many cases is contained in the behavioural equivalence relation _=b=_, and therefore is available for coinduction proofs (see Section 14). This is a hidden relation between the states whose "attributes" evaluate to the same values of data, i.e.,

$$a =*= a' \text{ iff } \sigma(d_1, \ldots, d_n, a) = \sigma(d_1, \ldots, d_n, a')$$

for $\sigma \in \Sigma_{v_1 \ldots v_n h, v}$ with $v_1, \ldots, v_n, v \in V$ and $h \in H$, $d_i \in D_{v_i}$ are values of data of sort v_i for $i \in [n]$. (We assume here the argument of hidden sort comes in the last place.)

The coinduction method requires that _=*=_ is a congruence; when this is the case, then _=*=_ is the behavioural equivalence. For example, this holds whenever the specification contains equations

$$attr(meth(x)) = t'$$

for all possible combinations of attributes *attr* and methods *meth*, and t' is a term built only of attribute or data operations. This situation appears quite frequently since it basically corresponds to describing the new state after the application of a method.

Example 15 Consider the following behavioural specification of "bags with natural numbers as elements":

```
mod! BARE-NATP {
  protecting(BARE-NAT)

  op p_ : Nat -> Nat

  eq p 0 = 0 .
  eq p s N:Nat = N .
}


mod* BAG {
  protecting(BARE-NATP)
```

```
[ Elt ]    *[ Bag ]*

op empty :    -> Bag
bop put : Elt Bag -> Bag
bop take : Elt Bag -> Bag
bop get : Bag Elt -> Nat

vars E E' : Elt
var B : Bag

eq get(empty, E) = 0 .
cq get(put(E, B), E')   =   get(B, E') if E =/= E' .
eq get(put(E, B), E)    = s(get(B, E)) .
cq get(take(E, B), E')  =   get(B, E') if E =/= E' .
eq get(take(E, B), E)   = p(get(B, E)) .
}
```

In this specification, bags have only one "attribute", get, and two "methods",
put and take. The equations express completely the change of the attribute
after an application of each of the two methods. Therefore, =*= is the behavioural equivalence.

Alternatively, we might specify bags with "methods" as non-behavioural
operations:

```
mod* BAG {
  protecting(BARE-NATP)

  [ Elt ]   *[ Bag ]*

  op empty :    -> Bag
  op put : Elt Bag -> Bag    {coherent}
  op take : Elt Bag -> Bag   {coherent}
  bop get : Bag Elt -> Nat
-- exactly the same as previous BAG
.

.

.

}
```

by specifying `put` and `take` as non-behavioural operations. Notice that both `put` and `take` are in fact *behaviourally coherent*. This specification is essentially equivalent to the previous one, but expresses directly (i.e., without needing proofs or arguments to justify it) the basic insight that the behavioural equivalence for bags is just `=*=`. □

Other examples involving `=*=` (also showing the support provided by CafeOBJ) can be found in Section 16.

Finally, the following table shows a comparative summary of the CafeOBJ built-in semantic relations.

Paradigm	equational	rewrite	behavioural
Syntactic Symbol	= (eq)	=>	= (beq)
Semantic built-in Relation	_==_	_==>_	_=b=_

5   Proof System

The semantics of algebraic languages in general, and of CafeOBJ in particular should be understood at three different levels:

- denotational semantics,
- proof-theoretic semantics (proof calculus), and
- operational semantics.

The correctness of the execution of the language is ultimately explained by the completeness of some operational semantics. It is also important to understand that while the denotational level should always be uniquely defined, there may be various operational semantics for the same language (even though only one may be thought as the default one). The completeness of the operational semantics with respect to the denotational semantics is a two-layer completeness since it can be decomposed as the completeness of the operational semantics with respect to proof calculus and the completeness of the proof calculus with respect to the denotational semantics.

We denote the deduction relation induced by the proof calculus by $\vdash$. Then, the completeness of the proof calculus with respect to the denotational semantics means that given a specification SP and a sentence ρ, then

$$SP \models \rho \text{ if and only if } SP \vdash \rho$$

In CafeOBJ both equational and rewriting specification have their own complete proof calculi. By contrast, behavioural specification does not admit a complete set of inference rules; however advanced proof techniques for behavioural specification are supported in CafeOBJ (see Section 14).

We present proof calculi deriving uncoditional sentences, this style being closer to the operational semantics of CafeOBJand sometimes referred as *local* proof calculi [6]. For a more *global* style of proof calculi deriving conditional sentences see [60].

5.1 Equational logic proof calculus

We present the equational logic proof calculus with respect to a fixed equational specification SP. This consists of the following inference rules:

[reflexivity] $\dfrac{}{(\forall X)\, t \;=\; t}$

[symmetry] $\dfrac{(\forall X)\, t \;=\; t'}{(\forall X)\, t' \;=\; t}$

[transitivity] $\dfrac{(\forall X)\, t \;=\; t' \quad (\forall X)\, t' \;=\; t''}{(\forall X)\, t \;=\; t''}$

[congruence] $\dfrac{(\forall X)\, t_i \;=\; t'_i \ \text{ for } \ i \in [n]}{(\forall X)\, \sigma(t_1,\ldots,t_n) \;=\; \sigma(t'_1,\ldots,t'_n)}$

for all operations $\sigma \in sign(SP)_{s_1\ldots s_n s}$, and t_i of sort s_i for $i \in [n]$.

[substitutivity] $\dfrac{(\forall X)\, \theta(C) \;=\; \texttt{true}}{(\forall X)\, \theta(t) \;=\; \theta(t')}$

where $(\forall Y)\, t \;=\; t'$ **if** C is any equation in SP and $\theta : Y \to T_{sign(SP)}(X)$ is any substitution.

Completeness of equational deduction constitutes one of the fundamental results in algebraic specification and related research has a long tradition beginning with Birkhoff [2], continuing with the completenss of many-sorted equational deduction [34], of order-sorted equational deduction [38], and ending with completeness results for category-based equational logics [9]. Since the CafeOBJ conditional equations are slightly more general than the OSA conditional equations, the completeness of OSA deduction explains only partially the completeness of the equational proof calculus in CafeOBJ. For a full justification we refer to the completeness of constraint logic deduction [11, 12] which is an instance of the completeness of category-based equational deduction [9].

5.2   Rewriting logic proof calculus

The inference rules of the rewriting logic proof calculus are just the same as the rules for equational logic minus the symmetry rule.

[reflexivity] $$\frac{}{(\forall X)\, t \; => \; t}$$

[transitivity] $$\frac{(\forall X)\, t \; => \; t' \quad (\forall X)\, t' \; => \; t''}{(\forall X)\, t \; => \; t''}$$

[congruence] $$\frac{(\forall X)\, t_i \; => \; t_i' \;\; \text{for} \;\; i \in [n]}{(\forall X)\, \sigma(t_1,\ldots,t_n) \; => \; \sigma(t_1',\ldots,t_n')}$$

for all operations $\sigma \in sign(SP)_{s_1 \ldots s_n s}$, and t_i of sort s_i for $i \in [n]$.

[substitutivity] $$\frac{(\forall X)\, \theta(C) \; = \; \textbf{true}}{(\forall X)\, \theta(t) \; => \; \theta(t')}$$

where $(\forall Y)\, t \; => \; t'$ **if** C is any transition in SP and $\theta: Y \to T_{sign(SP)}(X)$ is any substitution.

If the specification *SP* includes a set of equations *E*, then we need to apply the previous rules *modulo* the equations *E*, which is just equivalent to adding one more rule to the proof calculus:

$$[\text{modulo}] \quad \frac{(\forall X)\, \theta(t) \;=> \; \theta(u)}{(\forall X)\, \theta(t') \;=> \; \theta(u')}$$

whenever $E \vdash (\forall X)\, t = t'$ and $E \vdash (\forall X)\, u = u'$.

For the completeness of the rewriting logic proof calculus we refer to Meseguer's seminal paper [49], with the proviso of the difference given by the difference between CafeOBJ conditional transitions and the original rewriting logic concept of conditional transition (rule).

5.3 Hidden algebra proof calculus

As mentioned above, hidden (sorted) algebra does not admit a complete set of inferece rules as proof calculus. However, it is important to mention that ordinary equational inference rules presented in Section 5.1 provide a *sound* inference system for behavioural equality [31, 33] *if and only if all non-behavioural operations on hidden sorts are behaviourally coherent.* CafeOBJ exploits this via the command breduce.

II Structuring Specifications

Besides basic specification constructs, CafeOBJ specifications include also structuring constructs. Structuring constructs constitute one of the most important concepts in modern software engineering since they give the possibility of systematic reuse of already written *modules* (i.e., labeled specifications including both structuring constructs and basic specification constructs) or libraries. CafeOBJ module system is based on the updating of the principles of the Clear [4, 5] and OBJ [22, 40] to multi-paradigm systems.

The CafeOBJ specification structuring operations act uniformly across the various CafeOBJ paradigms. The fundamental mathematical concept underlying the semantics of CafeOBJ modularization system is that of **extra theory morphism** [15, 14]. This extends the Clear-OBJ modularization approach based on *theory morphisms* (see [16]) to multi-paradigm systems. For a deeper understanding of the concepts underlying the CafeOBJ modularization system we recommend the reader study institutions and work on modularization using institutions in the Clear-OBJ tradition [29, 16, 14]. For a brief presentations of the main institution concepts and results related to the CafeOBJ specification structuring see the Institutions Appendix.

6   Fundamental Semantics Concepts

This section is devoted to the brief presentation of some mathematical concepts underlying the semantics of structured specifications.

Signature Morphisms

Given CafeOBJ signatures $(S, \leq, \Sigma)$ and $(S', \leq', \Sigma')$, then a **signature morphism** $\phi \colon (S, \leq, \Sigma) \to (S', \leq', \Sigma')$ consists of

- a mapping of sorts $f \colon S \to S'$ such that $f(s) \leq' f(s')$ if $s < s'$, and
- an indexed family of mappings on operations, i.e.,

$$\{g_{s_1 \ldots s_n s} \colon \Sigma_{s_1 \ldots s_n s} \to \Sigma'_{f(s_1) \ldots f(s_n) f(s)}\}_{s_1, \ldots, s_n, s \in S, n \geq 0}$$

satisfying the following conditions:

1. f maps each hidden sort to a hidden sort,
2. if $f(h) <' f(h')$ for any hidden sorts h, h', then $h < h'$,
3. g maps each behavioural operation to a behavioural operation, and
4. if $\sigma' \in \Sigma'_{w's'}$ is a behavioural operation and some sort in w' is hidden, then $\sigma' = g(\sigma)$ for some behavioural operation σ in Σ.

In behavioural specification, when dealing with refinement of specifications, the last condition is not necessary. Signature morphism without 4. are called **vertical**, while ordinary signature morphism are called **horizontal** signature morphisms, suggesting their "orthogonal" use, i.e., refinement vs. module composition.

Signature morphisms can be **composed**, i.e., given signature morphisms $(f, g) \colon (S, \leq, \Sigma) \to (S', \leq', \Sigma')$ and $(f', g') \colon (S', \leq', \Sigma') \to (S'', \leq'', \Sigma'')$, then $(f; f', g; g')$ is a signature morphism $(S, \leq, \Sigma) \to (S'', \leq'', \Sigma'')$, where

- $(f; f')(s) = f'(f(s))$ for each sort $s \in S$, and
- $(g; g')_{s_1 \ldots s_n s}(\sigma) = g'_{f(s_1) \ldots f(s_n) f(s)}(g_{s_1 \ldots s_n s}(\sigma))$ for all $s_1, \ldots, s_n, s \in S$ and $n \geq 0$ and for all $\sigma \in \Sigma_{s_1 \ldots s_n s}$.

This definition applies both to horizontal and vertical signature morphisms, and composition of signature morphisms is associative and has left and right identities.

Sentence Translations

Each CafeOBJ signature morphism $\phi = (f,g)\colon (S,\leq,\Sigma) \to (S',\leq',\Sigma')$ induces a translation of CafeOBJ $(S,\leq,\Sigma)$-sentences to $(S',\leq',\Sigma')$-sentences (denoted by ϕ too) in the following way:

- the sort s of each variable is changed to the sort $f(s)$,
- each Σ-term $\sigma(t_1,\ldots,t_n)$ is translated to the Σ'-term $g(\sigma)(\phi(t_1),\ldots,\phi(t_n))$,
- each Σ-equation $(\forall X)\, t = t'$ **if** C is translated to $(\forall \phi(X))\, \phi(t) = \phi(t')$ **if** $\phi(C)$,
- the same for transitions, and
- behavioural sentences are translated to behavioural sentences.

Specification Morphisms

In this Report we use "specification morphism" as terminology rather than "theory morphism". A **specification morphism** $\phi\colon SP \to SP'$ is a signature morphism $\phi\colon sign(SP) \to sign(SP')$ such that

$$SP' \models \phi(\rho) \quad \text{for each } \rho \in SP$$

where SP and SP' are finite "specifications", i.e., finite collections of sentences for the corresponding signatures. This definition includes structured specifications by "flattening" them to basic specifications.[1] In fact, we will very often identify structured specifications with their corresponding basic specification obtained by "flattening".

Notice that given a specification morphism $\phi\colon SP \to SP'$, the paradigm underlying SP should always be embedded in the paradigm underlying SP' (in other words there should be a CafeOBJ cube arrow between the underlying institutions). Specification morphisms can be **composed** just as signature morphisms.

---

[1] We leave the precise definition of the "flattening" of structured specifications to basic specifications to the reader. For example, this can be done by defining the flattening for each structuring construct. In the SRA implementation, the command `describe` shows the flattened module.

In CafeOBJ the language construct implementing specification morphisms is **view**.

Model Reducts

Given a specification morphism $\phi = (f, g) \colon SP \to SP'$, then each model M' of SP' can be "reduced" to a model M of SP. This corresponds to "extracting" an implementation for SP from each implementation of SP', and can be formally defined by

(RD1.1) $M_s = M'_{f(s)}$ for each sort s of SP,

(RD1.2) $M_\sigma = M'_{g(\sigma)}$ for each operation σ of SP'.

When SP' is a rewrite specification (possibly behavioural too) and SP is not, then another reduction step is necessary for getting an algebra by "forgetting" the rewrite structure of M.

(RD2) all transitions from the carriers of M are removed, and consequently, all operations of M are reduced from functors to functions.

Model reducts realize a simplification of structure. While the steps (RD1.*) realise this within a fixed paradigm, (RD2) simplifies across paradigms, i.e., from a more complex paradigm to a simpler one. In the case of CafeOBJ results on model reducts for extra theory morphisms developed at the level of institutions [15, 14] assure the interchangeability of the two reduction steps. This is due to the good properties of the relationships between the CafeOBJ paradigms/logics (i.e., the CafeOBJ cube arrows being institution embeddings rather than simple institution morphism; see [15, 14] for details). Given a specification morphism $\phi \colon SP \to SP'$, then the reduct of a model M' of SP' is denoted as $M'\!\restriction_\phi$.

The following example involves an (intra) reduct just within the OSA institution, while the next one involves also a reduct across the institution embbeding OSA $\longrightarrow$ OSRWL (see the CafeOBJ cube).

Example 16 Consider the (inclusion) signature morphism
φ: *sign*(BARE-NAT) $\hookrightarrow$ *sign*(STRG-BNAT) of Example 11. Then

     StrgNat$\restriction_\varphi$ = **Nat**

by "forgetting" the interpretation of the sort Strg, and the interpretations of
nil. $\square$

Example 17 Consider an enrichment of the unreliable strings Example 13
with a length operation:

```
mod! USTRG#-BNAT {
   protecting(USTRG-BNAT * { sort Nat -> Number })
   protecting(BARE-NAT)

   op #_ : Strg -> Nat

   eq # nil = 0 .
   eq # (N:Number . S:Strg) = s(# S) .
}
```

Notice that we rename the sort Nat of USTRG-BNAT in order to distinguish
(separate) the elements of the strings from the results of the length function.
     Then we can expand the RWL model **UStrgNat** to a model **UStrgNat#**
interpreting # as the length of strings and interpreting Nat as natural numbers
with exactly one transition from a bigger one to a smaller one, and interpreting
Number as natural numbers without transitions. Notice that this is *not* the
initial model, since in the initial model there are infinite number of transitions
between the natural numbers of Nat, generated by the application of the length
functor to the transitions between strings.[2]
     The reducts along φ': *sign*(STRG-BNAT) $\hookrightarrow$ *sign*(USTRG#-BNAT) in-
volve both an "extra" reduction (forgetting the transitions between strings and
numbers) and an "intra" reduction (forgetting the interpretation of # and of
Nat with its adjacent operations). Notice that the source signature should be

---

[2]This shows some inadequacy for RWL specification; one way to avoid this would be to
interpret only the "constructors" as functors, and interpret the rest of operations as functions.
However, since this Report advocates a different usage of RWL than for ordinary specification,
such extension lies outside the main scope of this Report.

thought as an OSA signature, while the target should be thought as an OSRWL signature. These two reduction steps are interchangeable, as seen from

$$\textbf{UStrgNat\#}\!\restriction_{\varphi'} = \textbf{StrgNat}$$

So, we may reduce **UStrgNat#** first to **UStrgNat**, or else, reduce it first to an OSA model **StrgNat#** (expanding **StrgNat** with the natural interpretation of the length operation with values in a distinct copy of the naturals) as in the following figure:

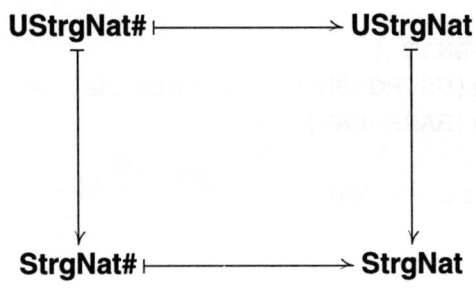

$\square$

The "Satisfaction Condition"

The so-called **Satisfaction Condition** of institutions theory [29, 16] provides the foundations for the correctness of specification structuring. This is:

$$M'\!\restriction_{\phi} \models \rho \ \text{ iff } \ M' \models \phi(\rho)$$

where $\phi\colon \Sigma \to \Sigma'$ is a signature morphism, M' is any Σ'-model, and ρ is any Σ-sentence. A direct consequence of the Satisfaction Condition is the following more software engineering oriented condition:

$$M'\!\restriction_{\phi} \models SP \ \text{ if } \ M' \models SP'$$

where $\phi\colon SP \to SP'$ is a specification morphism. This condition also expresses the correctness of the reduction process along a specification morphism. In the case of CafeOBJ, again due to its multi-paradigm nature, the above property incorporates both a Satisfaction Condition local to a fixed institution, and a

Satisfaction Condition for institution morphisms (see [29, 15, 14] for the details). This relies on the Satisfaction Condition for each of the particular institutions and each of the institution morphisms in the CafeOBJ cube (consult the basic references in the paradigm table in the Introduction for more details on the Satisfaction Condition for each specific institution).

Model Expansions

The concept of **model expansion** is dual to model reducts, and plays a crucial rôle for defining the denotations of structured specifications.

Given a specification morphism $\varphi \colon SP \to SP'$, and a model M of SP, an **expansion of M along** φ is a model M' of SP' satisfying the following properties:

- $M' \restriction_\varphi = M$ iff the expansion is *protecting*,
- there is an *inclusive* model morphism $M \hookrightarrow M' \restriction_\varphi$ (i.e., M is a *sub-model* of $M' \restriction_\varphi$ iff the expansion is *extending*,
- there is an arbitrary model morphism $M \to M' \restriction_\varphi$ iff the expansion is *using*, and
- there is a morphism $\eta_M \colon M \to M' \restriction_\varphi$ such that for any other SP'-model N' and morphism $h \colon M \to N' \restriction_\varphi$ there exists a unique morphism $h' \colon M' \to N'$ with $h = \eta_M; h' \restriction_\varphi$ (see the diagram below) iff the expansion is **free** .

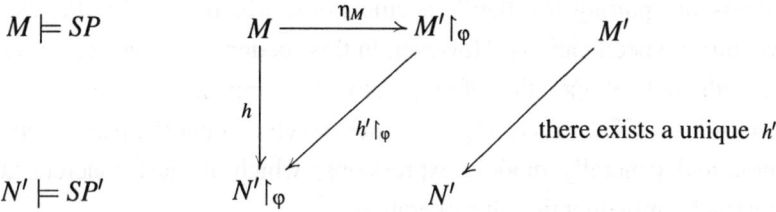

$$M \models SP \qquad\qquad M \xrightarrow{\ \eta_M\ } M' \restriction_\varphi \qquad\qquad M'$$

$$N' \models SP' \qquad\qquad N' \restriction_\varphi \qquad\qquad N'$$

While *protecting*, *extending*, and *using* concepts of model expansions are used for defining structuring modes (such as importation modes), free expansions play a crucial rôle in the semantics of module imports and of parameterization by generalizing the famous ADJ *initial algebra semantics* [39], the corresponding category theory notion being that of **adjoint functor** (see [45]).

In institution terminology, the property of free expansions of models is called **liberality** ([29, 16], see also the Appendix on institutions). In the case of CafeOBJ, due to its multi-paradigm nature, we have to deal with a more general form of liberality across the CafeOBJ cube. The basic mathematical results supporting free extensions for the CafeOBJ framework can be found in [14].

Example 18 In Example 16, **StrgNat** is a free *protecting* expansion of **Nat** along the inclusion specification morphism BARE-NAT $\hookrightarrow$ STRG-BNAT. Also, in Example 17, **UStrgNat** is a free *protecting* expansion of **StrgNat** along STRG-BNAT $\hookrightarrow$ USTRG-BNAT and **UStrgNat#** is a *protecting* but *not* free (since Nat is not interpreted freely) expansion **UStrgNat** along USTRG-BNAT $\hookrightarrow$ USTRG#-BNAT.

In all these cases the freeness of the expansion models is a direct consequence of their initiality property. $\square$

Model Amalgamation

The property of **model amalgamation** expresses the possibility to put together implementations which are consistent to each other into a big implementation. In the theory of institutions, model amalgamation is known also under the name of **exactness** [16, 14], and its rôle for modularization was first studied extensively in [16].

In order to discuss model amalgamation, we first need to discuss the basic process of "putting together" specifications, which is in fact the essence of structuring specifications. However, in this section, we concentrate on the basic mathematical operation of putting together implementations, rather than on various CafeOBJ structuring constructs (such as imports, parameters, views, sums, and generally module expressions) which are just concrete language constructs implementing this operation.

CafeOBJ follows the tradition of the OBJ family of languages in which specification structuring is based on the categorical concept of **co-limit** (see [5, 29, 16]). Because complex co-limits can be obtained as iterations of pushouts (a very simple form of co-limit)[3], we will concentrate here on pushouts. Be-

---

[3]This is in fact one of the fundamental theorems in basic category theory [45].

cause of their crucial rôle in structuring specifications, pushouts are now part
of the folklore of the theory of algebraic specification. All basic structuring
concepts, such as multiple imports (with sharing), parameterization, sums, can
be modeled as pushouts in a simple direct way.

Given two specification morphisms $\phi\colon SP \to SP_1$ and $\varphi\colon SP \to SP_2$, their
pushout consists of a specification SP' and two morphisms $\phi'\colon SP_2 \to SP'$ and
$\varphi'\colon SP_1 \to SP'$ such that $\phi;\varphi' = \varphi;\phi'$ and such that they are the "minimal"ones
with this property

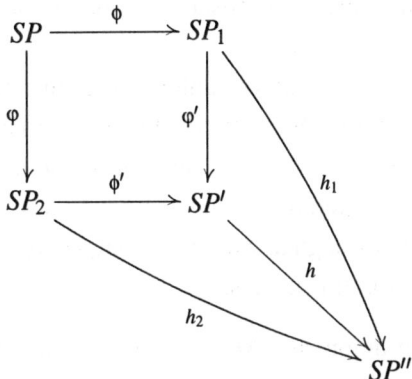

i.e., for all specification morphisms h_1, h_2 with $\varphi;h_2 = \phi;h_1$ there exists an
unique specification morphism h such that $\phi';h = h_2$ and $\varphi';h = h_1$.

Pushouts are unique up to isomorphisms (i.e., given ϕ and φ, any two
pushouts are essentially the same). Pushouts of specifications can be constructed in two distinct steps:

- first construct the pushout of signatures,

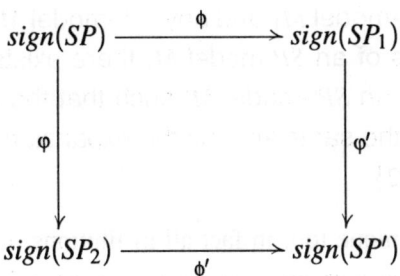

- then define the sentences of SP' to be the union of $\varphi'(SP_1)$-sentences and $\phi'(SP_2)$-sentences, i.e., $\varphi'(SP_1) \cup \phi'(SP_2)$.

These two steps are justified by the basic result on co-limits of theories for institutions (see [29] for the "intra" version and [14] for the "extra" version along institution morphisms). A special mention should be made for the case of pushouts of order-sorted signatures which can raise some non-trivial problems, especially in connection to the preservation by the pushout of some practically important properties of the signatures, such as the so-called *regularity* [38, 30]. Some results supporting pushouts of order sorted signatures can be found in [41].

Intuitively, pushouts are a kind of union, where the rôle of the intersection is played by the shared part. More concretely, the pushout of signatures can be obtained as a disjoint union of the two signatures, and then by identifying the elements (either sorts or operations) which are the image of the same element from the "shared" part (through any of the two signature morphisms involved). With the notations above, this means

- take the disjoint union $sign(SP_1) \cup sign(SP_2)$, and then
- identify (collapse) the elements $\phi(z)$ and $\varphi(z)$ for each $z \in sign(SP)$; this is $sign(SP')$.
- ϕ' and φ' are the natural mappings from SP_2 and, SP_1 respectively, which first inject the elements into the disjoint union, and then possibly identify them.

The amalgamation property means that given such a pushout of specifications,

> for any SP_1-model M_1 and any SP_2-model M_2 which are both expansions of an SP-model M, there exists a common expansion to an SP'-model M' such that the expansion of M_i to M' is of the same kind as the expansion of M to M_j, with $\{i, j\} = \{1, 2\}$.

In general M' is unique too; in fact all institutions of interest support unique amalgamation. However, amalgamation property might fail across institution morphisms (see [14]). Fortunately, there are sufficient conditions on the corresponding institution morphisms such that in most cases (covering all practical situations) amalgamation property holds (see [14]).

The most common case is when all the expansions involved are *protecting*; this is also the best known and most extensively studied case in the algebraic specification literature. However, cases involving a pair of *extending* and a pair of *protecting* expansions may be quite frequent (see Example 23).

Denotations of Structured Specifications

Structured specifications can be regarded as specifications constructed from basic specifications by iterations of structuring constructs. Therefore,

> The denotations of structured specifications (i.e., the class of models they denote) can be obtained from the denotations of basic specifications by iterations of semantic constructs corresponding to the structuring constructs of the specification.

The semantics of structuring CafeOBJ specifications is nothing more than defining the denotations of structured specification; how to do this will become gradually clear over this part.

As in the case of basic specification, at the top level, structured specifications may have two kinds of denotations: **tight** and **loose**. These are supported syntactically exactly in the same way as for basic specifications, i.e., by mod!, and mod* respectively. However, the actual denotations of structured specifications lie in general between the tight and the loose extremes.

We also extend the notation $[\![\_]\!]$ from basic specifications to structured specifications.

7   Module Imports

Module imports are the CafeOBJ fundamental structuring specification mechanism. A **module import** is a (basic) specification morphism $SP \to SP'$ *which is an inclusion* at the level of the signatures, i.e., $sign(SP) \hookrightarrow sign(SP')$. Then

SP is referred as the **imported module**, and *SP'* as the **importing module**. We denote module imports by $\unlhd$ and strict module imports by $\lhd$.

Notice that module imports form a partial order (i.e., there is at most one import between two given specifications), hence we can simplify the notation of module reducts for module imports $SP \unlhd SP'$ to $\_\!\restriction_{SP}$ because we do not need any name for the specification morphism defining the module import.

7.1   Importation Modes

There are 3 **importation modes** accordingly to the semantic classification, and in CafeOBJ they are declared explicitly: ***protecting, extending***, and ***using***. By convention we refer to *protecting* as the strongest and to *using* as the weakest. These declarations are semantic, and they *determine the denotations (semantics) of structured specifications* but no complete checking support should be expected from the system. However, there are some sufficient syntactic conditions ensuring the consistency of the importation modes declarations. In order to avoid inconsistent specifications (i.e., specifications without any models), it is important to declare accurately the importation mode one uses.

The following defines the denotation of imports. Fix an import $SP \unlhd SP'$. Then $[\![SP']\!]$ is defined as

- $\{M' \mid M' \models SP'$ and M' is *protecting* (and free if SP' has initial denotation) expansion of some model $M \in [\![SP]\!]\}$, when the importation mode is *protecting*,

- $\{M' \mid M' \models SP'$ and M' is *extending* (and free if SP' has initial denotation) expansion of some model $M \in [\![SP]\!]\}$, when the importation mode is *extending*, and

- $\{M' \mid M' \models SP'$ and M' is *using* (and free if SP' has initial denotation) expansion of some model $M \in [\![SP]\!]\}$, when the importation mode is *using*, and

Protecting imports do not either collapse elements (or transitions) nor add new elements (or transitions) to the models of the imported module, but *extending* imports may add new elements but not collapse elements. In the folklore of algebraic specification these conditions are known under the name of "no junk and no confusion" and, "no confusion" conditions, respectively. Using imports provide no guaranty, so they might even collapse elements. Notice also that by "free" we mean that in each case the "universal" morphism η_M is actually the expansion morphism.

Example 19 In Example 11, the import BARE-NAT $\trianglelefteq$ STRG-BNAT is *protecting*, hence the denotation $[\![\text{STRG-BNAT}]\!]$ consists only of the initial model **StrgNat** because both BARE-NAT and STRG-BNAT are declared with tight semantics.

This example indicates a general pattern: whenever both the importing and the imported module have tight semantics and the denotation of the imported module consists only of the initial model, then the only candidate for the denotation of the importing module is the initial model too. This situation covers most of the cases when specifying data types, including STRG-BNAT $\trianglelefteq$ USTRG-BNAT discussed in Example 18. $\square$

Example 20 However, there is no guaranty that in similar cases (with the previous example) the *protecting* mode is respected. For example, consider the following "closure" of the natural numbers with an "infinity" element by importing BARE-NAT of Example 1:

```
mod! NAT-OMEGA {
  extending(BARE-NAT)

  op omega :    -> Nat
  pred _<=_ : Nat Nat

  vars N M : Nat

  eq 0 <= s(N) = true .
  cq s(M) <= s(N) = true    if M <= N .

  eq s(omega) = omega .
  eq N <= omega = true .
}
```

The free expansion of **Nat** is the initial model **Omega** of NAT-OMEGA which extends the carrier **Nat**$_\text{Nat}$ with a new element "infinity", i.e., **Omega**$_\text{Nat} = \omega \cup \{\omega\}$, which is a fixed point for the successor function. The new predicate _<=_ is just the usual total order on natural numbers with the infinity element as top element. Therefore, **Omega**⌈$_\text{BARE-NAT}$ is *not equal* to **Nat**, so in *protecting* mode the denotation ⟦NAT-OMEGA⟧ would be empty (no possible models). However, the unique morphism **Nat** → **Omega**⌈$_\text{BARE-NAT}$, is injective, hence the proper importation mode is *extending*.

Another example of an *extending* import (this time across the CafeOBJ cube) is given by the RWL specification of nondeterministic naturals (the nondeterministic aspect of it is discussed in Section 15.1; here we concentrate on the import aspect only):

```
mod! BARE-NNAT {
   extending(BARE-NAT)

   op _|_ : Nat Nat -> Nat {assoc}

   vars M N : Nat

   eq s(M | N) = s(M) | s(N) .

   trans M | N => N .
   trans M | N => M .
}
```

Here the carrier **NNat**$_\text{Nat}$ of the initial model **NNat** of BARE-NNAT extends the natural numbers with nondeterministic expressions of the form (which can also be thought as strings of natural numbers)

$$n_1 \mid n_2 \mid \ldots \mid n_k$$

with $k \in \omega$ and with transitions to sub-expressions (similar to the transitions between strings of the model **UStrgNat**. Therefore, the unique algebra morphism **Nat** ↪ **NNat**⌈$_\text{BARE-NAT}$ is injective but not an isomorphism. □

Example 21 In Example 14, the module HSS-BNAT has loose semantics and imports in *protecting* mode the module BARE-NAT with tight semantics.

Therefore, the denotation $[\![\text{HSS-BNAT}]\!]$ consists of all models M of HSS-BNAT such that

$$M\!\restriction_{\text{BARE-NAT}} = \textbf{Nat}$$

Notice that the models **List** and **Arr** introduced in Example 9 satisfy the above condition. This is a general pattern for behavioural specification, since they always have loose semantics and importing the data in *protecting* mode. □

In all above examples the imported module had tight semantics; this situation occurs very often when specifying data types or when importing data in behavioural specification. The following example concentrates on the case when the importing module has loose semantics, which is generally the case for other structuring constructs such as sum of modules (with loose semantics) or parameters.

Example 22 Consider the following specification of sets:

```
mod* TRIV { [ Elt ] }
```

Its denotation consists of all (plain) sets, since any of its models is just an interpretation of the sort Elt as a set.

We can use TRIV for specifying the algebraic structure of monoids by adding an associative operation with identity.

```
mod* MON {
  protecting(TRIV)

  op nil :   ->  Elt
  op _;_ : Elt Elt -> Elt {assoc idr: nil}
}
```

Because MON has loose semantics *protecting* TRIV, its denotation consists of all monoids.

Commutative monoids can be specified by adding a commutativity equation (as an attribute in order to avoid non-termination) to the specification of monoids:

```
mod* CMON {
  protecting(MON)

  op _;_ : Elt Elt -> Elt {comm}
}
```

The denotation [[CMON]] consists of all commutative monoids.

Just by replacing the loose semantics declaration with a tight semantics one we get a specification of strings as free monoids:

```
mod! STRG' {
  extending(TRIV)

  op nil :    -> Elt
  op _._ : Elt Elt -> Elt {assoc idr: nil}
}
```

The denotation [[STRG']] consists of all free monoids. Given an interpretation for Elt as a set, the corresponding model of STRG' is the free monoid generated by that set, i.e., the strings with elements from that set. Notice that in this case the strings appear as new elements of the sort Elt (as in the case of BARE-NAT), therefore the proper importation mode is *extending*.

However, we can specify strings *protecting* TRIV by using a super-sort as follows:

```
mod! STRG {
  protecting(TRIV)

  [ Elt < Strg ]

  op nil :    -> Strg
  op _._ : Strg Strg -> Strg {assoc idr: nil}
}
```

□

Compositionality

The most characteristic property of module imports is that they form a partial order. The transitivity property of the partial order of module imports means that imports may be composed, and this composition is associative with left and right identities (by reflexivity). The most important consequence of this is that for an imported module the importation path does not matter. This kind of composition is often referred as **horizontal composition of module imports**. The importation mode of the composition of two or several imports is determined by the following rule:

> The importation mode of the horizontal composition of several imports is the weakest mode of the composing imports.

7.2   Sums and Multiple Imports

Modules may import several modules at the same time. In this case the submodules which are shared between the importing modules are imported only once. Consider two imports $SP_1 \trianglelefteq SP$ and $SP_2 \trianglelefteq SP$. If we denote by $SP_1 \wedge SP_2$ the shared specification between SP_1 and SP_2, then we have the situation described by the following diagram:

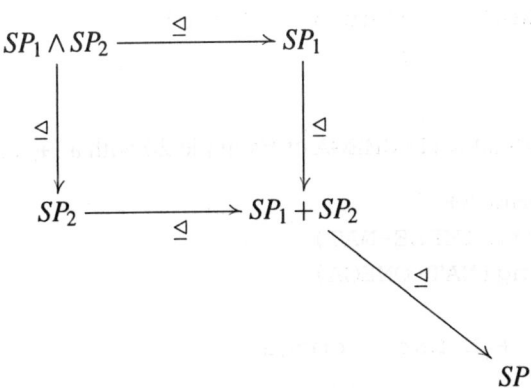

where $SP_1 + SP_2$ is the **(shared) sum** of SP_1 and SP_2. The shared sum $SP_1 + SP_2$ has several properties:

- it imports both SP_1 and SP_2, and
- the shared part $SP_1 \wedge SP_2$ is the greatest lower bound of SP_1 and SP_2 with respect to the partial order $\trianglelefteq$,

- it is the least upper bound of SP_1 and SP_2 with respect to the partial order $\trianglelefteq$, and the pushout of $SP_1 \wedge SP_2 \trianglelefteq SP_1$ and $SP_1 \wedge SP_2 \trianglelefteq SP_2$,
- $[\![SP_1 + SP_2]\!]$ consists of all models of $SP_1 + SP_2$ which are amalgamations of a model in $[\![SP_1]\!]$ with a model in $[\![SP_2]\!]$ over a model in $[\![SP_1 \wedge SP_2]\!]$.

The following rule determines the importation modes in case of sum of specifications:

> The import $SP_i \trianglelefteq SP_1 + SP_2$ has the same importation mode as $SP_1 \wedge SP_2 \trianglelefteq SP_j$ where $\{i, j\} = \{1, 2\}$.

Example 23 Consider the extension of BARE-NAT of Example 1 with a commutative _+_ operation:

```
mod! SIMPLE-NAT {
  protecting(BARE-NAT)

  op _+_ : Nat Nat -> Nat {comm}

  eq s(N:Nat) + M:Nat = s(N + M) .
  eq N:Nat + 0 = N .
}
```

Then we can extend NAT-OMEGA of Example 20 with a _+_ as follows:

```
mod! NAT-OMEGA+ {
  extending(SIMPLE-NAT)
  protecting(NAT-OMEGA)

  eq omega + N:Nat = omega .
}
```

which is equivalent to

```
mod! NAT-OMEGA+ {
  protecting(SIMPLE-NAT + NAT-OMEGA)

  eq omega + N:Nat = omega .
}
```

Notice that in this example NAT-OMEGA ⊴ NAT-OMEGA+ is in *protecting* mode, and SIMPLE-NAT ⊴ NAT-OMEGA+ is in *extending* mode. The situation of this example can be described by the following diagram:

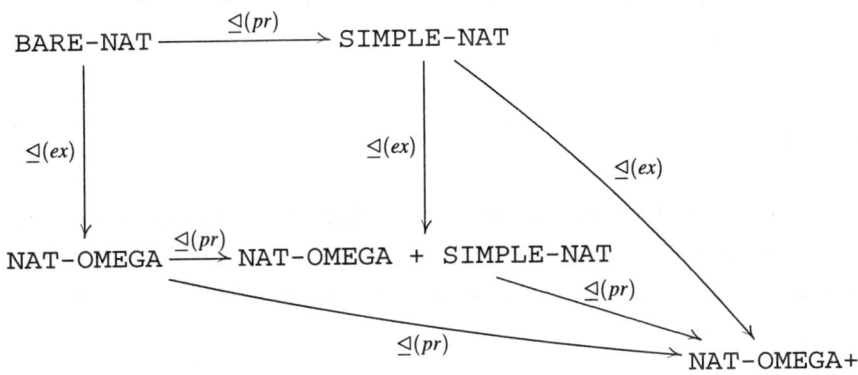

The denotation [[SIMPLE-NAT + NAT-OMEGA]] consists of all amalgamations M between **Omega** and **Nat+** over **Nat** (all these are the unique models of the corresponding denotations), where **Nat+** is the *protecting* expansion of **Nat** interpreting _+_ as addition of naturals. This means M is a *protecting* expansion of **Omega** with an interpretation of _+_ which is the standard one on the natural numbers, and arbitrary (but satisfying the axioms of _+_ of SIMPLE-NAT) on the "infinity". We can easily notice that in all amalgamations M, $n + M_{omega} = M_{omega} = M_{omega} + n$ for all $n \in M_{Nat}$ with $n \neq omega$, therefore the only difference between various amalgamations M would be the value of $M_{omega} + M_{omega}$. The only amalgamation M satisfying the equation of NAT-OMEGA+ is the one for which $M_{omega} + M_{omega} = M_{omega}$; let's denote it as **Omega+**. Hence **Omega+** is the denotation of NAT-OMEGA+. Notice that the *protecting* mode of the importation (NAT-OMEGA + SIMPLE-NAT) ⊴ NAT-OMEGA+ is respected too. □

Sum of specifications may also happen across the institution embeddings of the CafeOBJ cube. In this case the following rule applies:

The institution of $SP_1 + SP_2$ is the "least upper bound" of the institutions of SP_1 and SP_2 in the CafeOBJ cube.

Example 24 The sum NAT-OMEGA + BARE-NNAT defining the nondeterministic naturals with an "infinity" element is an OSRWL specification because NAT-OMEGA is an OSA specification and BARE-NNAT is a OSRWL specification. The shared part is BARE-NAT, and all imports involved are in *extending* mode. □

8 Views

As mentioned above, views constitute the CafeOBJ implementation of the mathematical concept of specification morphism. A **view** between specifications *SP* and *SP'* is a specification morphism with *sign(SP')* possibly being a derived signature.

Example 25 Here we discuss some examples of views. Consider the specification of monoids of Example 22. Then,

```
view bare-nat from TRIV to BARE-NAT {sort Elt -> Nat}
```

is a view selecting the set of naturals,

```
view plus from MON to SIMPLE-NAT {
   sort Elt -> Nat,
   op _;_ -> _+_,
   op nil -> 0
}
```

is an example of a view regarding the natural numbers as the additive monoid, and

```
view dual from MON to MON {
   op X:Elt ; Y:Elt -> Y:Elt ; X:Elt
}
```

takes the "dual" of a monoid.

A slightly more sophisticated view uses a derived operation on the natural numbers:

```
view star from MON to NAT {
   sort Elt -> Nat,
   op nil -> 0,
   op X:Elt ; Y:Elt -> X:Nat + Y:Nat + X * Y
}
```

where NAT is the module of built-in natural numbers and _*_ is the multiplication operation on natural numbers.

All these views are specification morphisms; for the first one the checking means the proof of inductive associativity of _+_, for the second one the checking is trivial, and the proof for star requires some calculations; here is its proof score:

```
ops x y z :   -> Nat .
op _;_ : Nat Nat -> Nat .
eq X:Nat ; Y:Nat = X + Y + X * Y .

red (x ; y) ; z == x ; (y ; z) .
red x ; 0 == x .
red 0 ; x == x .
```

□

9   Parameterized Modules

In CafeOBJ a **parameterized module (specification)** $SP(X :: P)$ can be regarded as a labeled *injective* (not collapsing elements of the signatures) specification morphism $P \xrightarrow{X} SP$, where P is the **parameter** specification, SP is the **body**, and X is the **label** (of the parameter; we often call the parameter by its label). Various instances of a parameterized module can be obtained by replacing the parameter(s) with other modules satisfying the constraints of the parameter. The satisfaction of the parameter by other modules means exactly a specification morphism (view). There may be different ways in which a module satisfies a parameter; this corresponds to different specification morphisms (views). Parameters can be regarded like a kind of generic imports, i.e., imports which can be further instantiated to other imports. In fact the image of the parameter via its label should be considered as an import, i.e., $X(P) \trianglelefteq SP$.

However, a fundamental difference between imports and parameters is that the latter obeys a principle of non-sharing (this will be explained later), hence they are labeled. This principle of non-sharing for parameters may have two versions: a strong one, and a weak one allowing the sharing of the imports. This amounts to two different approaches on parameterization: the **non-shared parameterization** and the **shared parameterization**. CafeOBJ supports both of them, and the latter can be conceptually regarded as a particular case of the former by imposing a special equation in the corresponding module calculus. Since non-shared parameterization is conceptually more general, we will develop the parameterization concepts for the non-shared approach, and then discuss the shared approach.

From a practical perspective, non-shared parameterization gets a more powerful module system while the shared parameterization is easier to be efficiently implemented.

9.1 Non-shared Parameterization Modes

As in the case of module imports, there are three **parameterization modes**: *protecting*, *extending*, and *using*. The default parameterization mode is the *protecting* one. Depending on the actual parameterization mode, the **denotation of a parameterized module** $SP(X :: P)$ (with only one parameter) can be defined similarly to the denotation of importing modules just by using expansions of models along the (injective) specification morphism X rather than along an inclusion $P \hookrightarrow SP$.

Example 26 The following are parameterized specifications of monoids, commutative monoids and of strings (recall the similar non-parameterized specifications of Example 22):

```
mod* MON* (X :: TRIV) {
-- the same operations and equations as MON

  op nil :   ->  Elt
  op _;_ : Elt Elt -> Elt {assoc idr: nil}
}
```

```
mod* CMON* (Y :: MON) {

   op _;_ : Elt Elt -> Elt {comm}
}

mod! STRG* (X :: TRIV) {
   [ Elt < Strg ]

   op nil :  -> Strg
   op _._ : Strg Strg -> Strg {assoc idr: nil}
}

mod! STRG'* (extending X :: TRIV) {

   op nil :  -> Elt
   op _._ : Elt Elt -> Elt {assoc idr: nil}
}
```

The denotation $[\![\text{MON}*]\!]$ consists of all monoids (i.e., the same as $[\![\text{MON}]\!]$), the denotation $[\![\text{CMON}*]\!]$ consists of all commutative monoids (i.e., the same as $[\![\text{CMON}]\!]$), the denotation $[\![\text{STRG}*]\!]$ consists of all string (free) monoids (i.e., the same as $[\![\text{STRG}]\!]$), and the denotation of $[\![\text{STRG}'*]\!]$ consists of all string free monoids too (i.e., the same as $[\![\text{STRG}']\!]$). Notice that the parameter $\text{TRIV} \xrightarrow{\text{X}} \text{STRG}'*$ is in *extending* mode. $\square$

Multiple Parameters

As with imports, parameterized modules may have **multiple parameters**. The fundamental difference between multiple imports and multiple parameters is that the former are shared, while the latter obey the following **principle of strong non-sharing**:

Let $P_1 \xrightarrow{X} SP$ and $P_2 \xrightarrow{Y} SP$ be two different parameters in the same parameterized module. Then the translations of P_1 and P_2 by X and Y respectively, are disjoint in SP (in more compact notation $X(P_1) \wedge Y(P_2) = \emptyset$). Also, translations of

parameters are disjoint from any module imports $SP' \triangleleft SP$
($X(P) \wedge SP' = \emptyset$ where $P \to SP$ is a parameter).

Example 27 Consider the (double parameterized) specification of a "power"
operation on monoids, where powers are elements of another monoid rather
than natural numbers.

```
mod* MON-POW (POWER :: MON, M :: MON)
{
  op _^_ : Elt.M Elt.POWER -> Elt.M

  vars m m' : Elt.M
  vars p p' :  Elt.POWER

  eq (m ; m')^ p   = (m ^ p) ; (m' ^ p) .
  eq  m ^ (p ; p') = (m ^ p) ; (m ^ p') .
  eq  m ^ nil      = nil .
}
```

The diagram defining MON-POW is

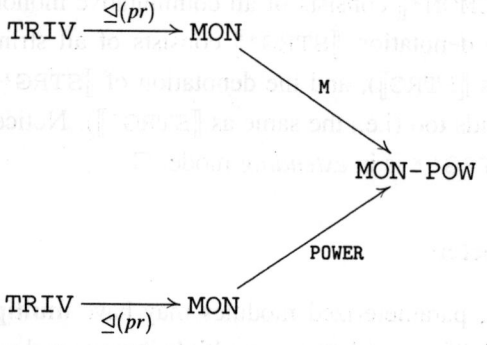

where MON-POW consists of two copies of MON (labeled by M and, POWER)
respectively, plus the power operation together with the 3 axioms defining its
action. This means TRIV is *not* shared, since the power monoid and the base
monoid are allowed to have different carriers.

The denotation ⟦MON-POW⟧ consists of all *protecting* expansions (with
interpretations of _^_) to MON-POW of non-shared amalgamations of monoids
corresponding to the two parameters.

The following specifies "free" powers of a monoid on another monoid:

```
mod! MON-POW! (POWER :: MON, extending M :: MON) {
-- exactly the same contents as MON-POW
  .

  .

  .
}
```

This specification is essentially the same with that of MON-POW with the difference that the parameterized specification has tight rather than loose semantics. Therefore, models of MON-POW! are *free* expansions of (non-shared) amalgamations of two different monoids. Given a monoid M^{POWER} for POWER and a monoid M^{M} for M, then the corresponding model M of MON-POW! is a *protecting* expansion of M^{POWER} along POWER ($M\lceil_{\text{POWER}} = M^{\text{POWER}}$), and an *extending* expansion of M^{M}. The carrier $M_{\text{Elt.M}}$ consists of all expressions of the form

$$(m_1 \hat{\ } p_1) ; \ldots ; (m_n \hat{\ } p_n)$$

where $n \in \omega$, $m_i \in M^{\text{M}}$ and $p_i \in M^{\text{POWER}}$ for all $i \in [n]$.

Notice that in this example, the same module appears as parameter in two different ways. However, if we specify commutative monoids with powers, the header would be

```
mod* CMON-POW(POWER :: MON, M :: CMON)
```

□

Importing parameterized modules

When parameterized modules are imported, the parameter becomes a parameter of the importing module, as can be visualized in the following diagram:

$$P \xrightarrow{\;\;X\;\;} SP \xrightarrow{\;\;\trianglelefteq\;\;} SP'$$

The parameterization mode of $SP'(X :: P)$ is the weakest between the importation mode of $SP \trianglelefteq SP'$ and the parameterization mode of $SP(X :: P)$. In general all concepts and principles of importation lift from the case of simple modules to the case of parameterized modules.

Parameterized parameters

A parameterized module can be itself a parameter for another module. For example, a situation like $SSP(X :: SP)$ with $SP(Y :: P)$ can be visualized as

$$P \xrightarrow{\;\;Y\;\;} SP \xrightarrow{\;\;X\;\;} SSP$$

In this case $Y;X$, which is also an injection can be regarded as (the label of) an **implicit parameter** of SSP.

Example 28 The specification of commutative monoids with powers (Example 27) uses the parameterized parameter CMON*. This example can be visualized as follows:

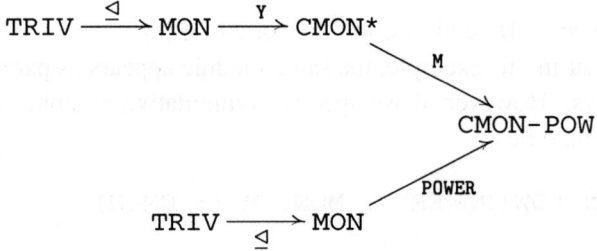

$\square$

9.2   Non-shared Parameter Instantiation

In this section we explain the basic mechanism of parameter instantiation within the non-shared framework.

The instantiation of $SP(X :: P)$ by a view $v: P \to P'$ is $SP(X<= v)$ (abbreviated as $SP(v)$ when the parameter is obvious, or even $SP(P')$ when the view is obvious too), and is defined by the following *pushout* of specification morphisms:

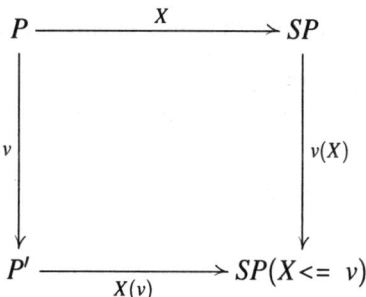

Instantiation of parameterized modules has several fundamental properties:

- the kind of denotation (i.e., tight or loose) of $SP(v)$ is the same with that of SP,
- the resulting specification morphism $X(v): P' \rightarrow SP(X<= v)$ is injective too,
- hence $SP(v)$ is a *parameterized* module $SP(v)(X(v) :: P')$,
- $X(v)$ preserves the parameterization mode of X, and
- the denotation $[\![SP(v)]\!]$ consists of all amalgamations of a model in $[\![SP]\!]$ with a model in $[\![P']\!]$ over a model in the denotation of parameter $[\![P]\!]$.

Example 29 The instantiation MON*(bare-nat) of the parameterized module MON* of Example 26 with bare-nat of Example 25 introduces an *arbitrary* monoid structure on the "bare" naturals. In order to map this abstract
monoid structure to a concrete one we can define a view:

```
view bare-nat+ from MON*(bare-nat) to SIMPLE-NAT {
    op _;_ -> _+_,
    op nil -> 0
}
```

□

It is important to notice that instances of parameterized modules are in
general different from the corresponding direct importation of the value of the
parameter. This is due to the fact that there is no sharing between different
parameters or between parameters and imports.

Example 30 Consider the following extension of the STRG* parameterized strings specification of Example 26 with a length operation.

```
mod! STRG#* (X :: TRIV) {
  protecting(STRG*(X) + SIMPLE-NAT)

  op #_ : Strg -> Nat

  eq # nil = 0 .
  eq # E:Elt = s 0 .
  eq # (S1:Strg . S2:Strg) = (# S1) + (# S2) .
}
```

The structure of the sub-modules of the instantiation STRG#* (bare-nat) is depicted in the following diagram:

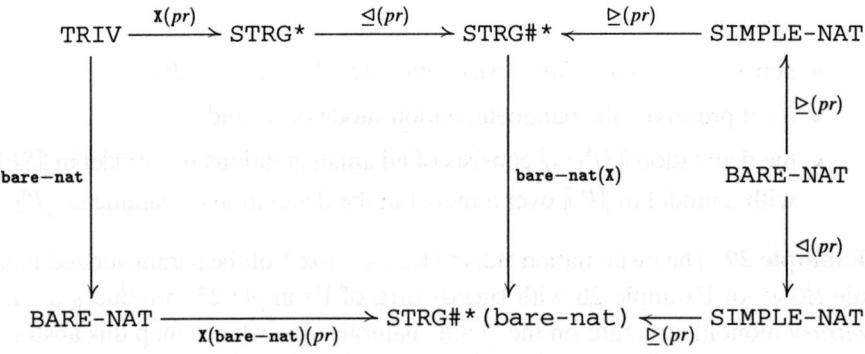

where the left hand side rectangle is a pushout. Notice that BARE-NAT is *not* shared since one is an import and the other one is a parameter (X(bare-nat)).

The situation is different for the following non-parameterized specification of the same problem:

```
mod! STRG#-BNAT {
  protecting(STRG-BNAT + SIMPLE-NAT)

  op #_ : Strg -> Nat

  eq # nil = 0 .
  eq # E:Nat = s 0 .
```

```
    eq # (S1:Strg . S2:Strg) = (# S1) + (# S2) .
}
```

In this case BARE-NAT is shared and the sub-modules structures looks like:

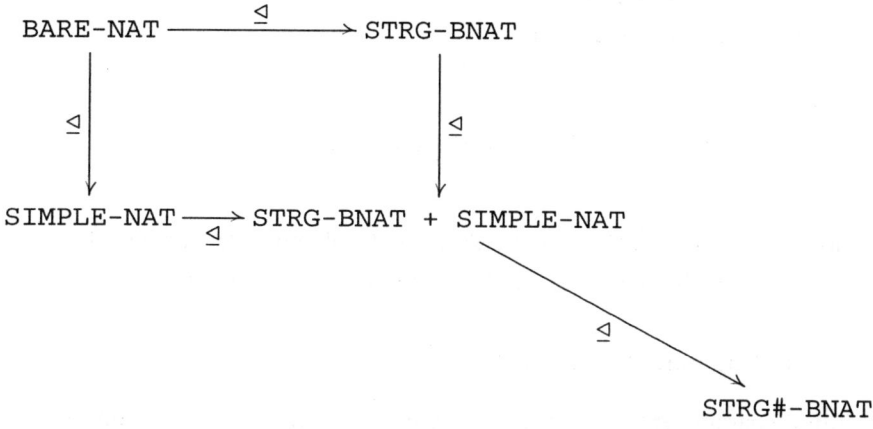

□

Parameters in *extending* mode should be used with great care (in general they should be avoided) since their instantiations might have eccentric denotations:

Example 31 Consider the instantiation STRG'*(bare-nat) of the *extending* parameterization STRG'*(extending X :: TRIV) of Example 26:

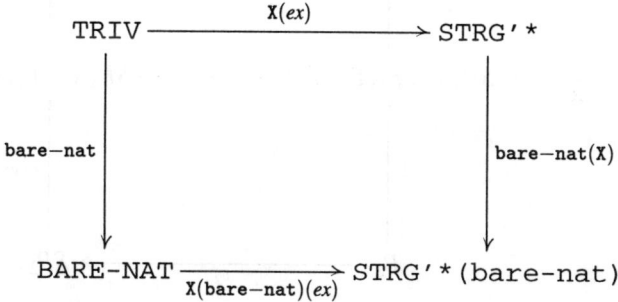

A model *M* in the denotation ⟦STRG'*(bare-nat)⟧ is just a *protecting* expansion (interpreting the successor operation s_) of the free monoid over the naturals ω. Because $s\_$ and $\_;\_$ do not interact at all, any function $\omega^* \to \omega^*$ will do this.

Therefore STRG'*(bare-nat) is different from the non-parameterized version of the same problem:

```
mod! STRG'-BNAT {
    extending(BARE-NAT)

    op nil :   ->  Nat
    op _;_  : Nat Nat -> Nat {assoc idr: nil}
}
```

whose denotation $[\![\text{STRG'-BNAT}]\!]$ consists of only one model which is a (ground) term algebra for $\_;\_$ and $s\_$ modulo the associativity of $\_;\_$. $\square$

Instantiating multiple parameters

Multiple parameters can be instantiated in any order, the instantiation steps being interchangeable. This is justified by basic mathematical results on pushout composition (see [45]), and can be visualized by the following commuting diagram:

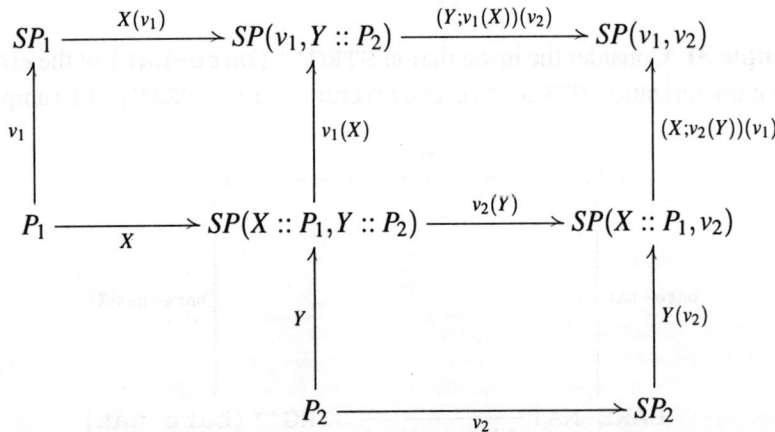

where all "atomic" squares are instantiation pushouts. Both rectangles formed each by two "atomic" squares are instantiation pushouts too (but of partial

instantiations). "Concurrent" instantiation of multiple parameters is equivalent
to taking the co-limit of

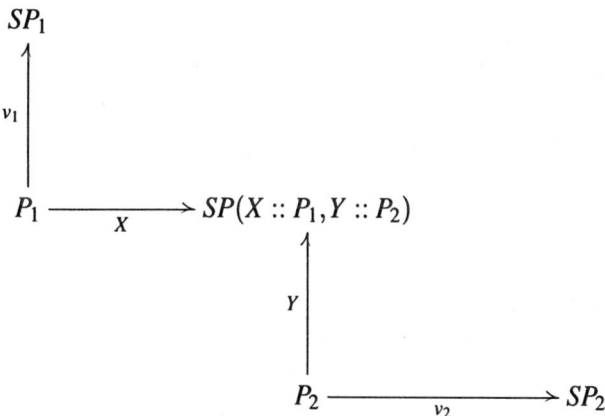

Notice also that both $X; v_2(Y)$ and $Y; v_1(X)$ are injective because X and Y are
not shared, hence they are proper parameters. Moreover, $v_1(X)$ and $v_2(Y)$
are identities on the images of X and Y, so we can abbreviate $X; v_2(Y)$ and
$Y; v_1(X)$ by X and, Y respectively.

Example 32 The doubly parameterized module MON-POW of Example 27
can be partially instantiated and gives a specification of monoids with natural powers:

```
mod* MON-POW-NAT {
  protecting(MON-POW(POWER <= plus))

  eq m:Elt.M ^ s(0) = m .
}
```

The other parameter can be instantiated too:

```
MON-POW(M <= plus)
```

These two steps can be interchanged or applied simultaneously for getting a
combined instantiation:

```
MON-POW(POWER <= plus, M <= plus)
```

Notice that the non-sharing principle implies that this instantiation contains two copies of the additive naturals. By collapsing these two copies of the naturals, we get the multiplication of naturals as the "power" of the addition:

```
mod* NAT-TIMES {
  protecting(MON-POW-NAT(M <= plus) *
    {sort Nat.M -> Nat, sort Nat.POWER -> Nat,
     op 0.M -> 0, op 0.POWER -> 0, op _^_ -> _*_}))
}
```

which is equivalent to

```
mod* NAT-TIMES {
  protecting(MON-POW(POWER <= plus, M <= plus) *
    {sort Nat.M -> Nat, sort Nat.POWER -> Nat,
     op 0.M -> 0, op 0.POWER -> 0, op _^_ -> _*_}))

  eq m:Nat * s(0) = m .
}
```

□

Instantiating parameterized parameters

The basic principle of instantiating a parameterized parameter in a module $SSP(X :: SP(Y :: P))$ by a view $v: SP \to SP'$ is that Y gets instantiated too. This of course includes the case when Y is instantiated to another parameter Y', as showed by the following diagram:

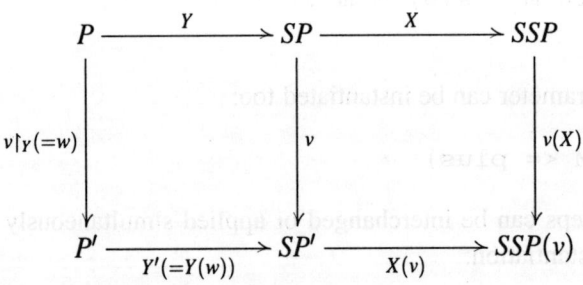

We distinguish several possible situations:

1. the view v preserves the parameter Y; in this case $P = P'$ and $Y' = Y$; v,
2. the view v explicitly maps the parameter Y to Y'; in this case a view $v\restriction_Y$ between the parameters should be given, and
3. the view v itself is the result of instantiating the parameter Y by a view w (meaning the left hand side square is a pushout too).

Instantiating across the CafeOBJ cube

When instantiating a parameterized module $SP(X :: P)$ with a view $v: P \to P'$, both X and v might be "extra" specification morphisms across some edge of the CafeOBJ cube. Then

> the paradigm (institution) of the result $SP(v)$ of the instantiation is the least upper bound of the paradigms (institutions) of SP and P' in the CafeOBJ cube.

Example 33 Consider the following parameterized behavioural specification of a counter:

```
mod* DATA {
  [ Data ]
  op _+_ : Data Data -> Data
}

mod* COUNTER(X :: DATA) {

  *[ Counter ]*

  bop add : Data Counter -> Counter
  bop read_ : Counter -> Data

  eq read add(N:Data, C:Counter) = read(C) + N .
}
```

We can then obtain a counter with natural numbers by instantiating the data to the natural numbers:

```
mod* COUNTER-NAT { protecting(COUNTER(SIMPLE-NAT)) }
```

But we may also obtain a nondeterministic counter by instantiating the data to the nondeterministic naturals:

```
mod! SIMPLE-NNAT {
  protecting(BARE-NNAT + SIMPLE-NAT)

  vars M N N' : Nat

  eq M + (N | N') = (M + N) | (M + N') .
}
```

```
mod* COUNTER-NNAT {protecting(COUNTER(SIMPLE-NNAT))}
```

The institution of COUNTER is HSA, the institution of SIMPLE-NAT is OSA, and the institution of SIMPLE-NNAT is OSRWL. Hence the institution of COUNTER-NAT is HOSA and the institution of COUNTER-NNAT is HOS-RWL. □

9.3 Shared Parameters

In this section we discuss parameter sharing in CafeOBJ. This is based on the following **weak non-sharing principle:**

> Let $P_1 \xrightarrow{X} SP$ and $P_2 \xrightarrow{Y} SP$ be two different parameters in the same parameterized module. Then the translations of P_1 and P_2 by X and Y respectively, are disjoint in SP *outside their common imports*; in mathematical notation:
>
> $$X(P_1) \wedge Y(P_2) = (\sum_{M \trianglelefteq P1} M) \wedge (\sum_{M \trianglelefteq P2} M)$$
>
> Similarly for the module imports of the body of the parameterized module:
>
> $$X(P) \wedge SP' = (\sum_{M \trianglelefteq P} M) \wedge SP'$$
>
> where $P \to SP$ is a parameter and for all $SP' \triangleleft SP$.

The diagram calculus for the shared parameterization can be obtained by adding the condition

$$M \xrightarrow{\triangleleft} P \xrightarrow{X} SP = M \xrightarrow{\triangleleft} SP$$

to the non-shared parameterization calculus, which means quotienting the corresponding diagrams modulo the commutativity of the following triangles:

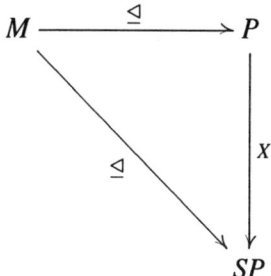

Example 34 Consider the specification MON-POW of a "power" operation on monoids (Example 27). Then the shared parameterization diagram corresponding to this example is

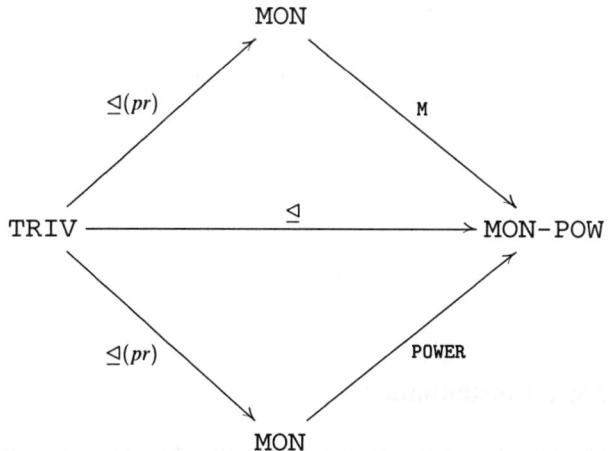

In this case, the denotation ⟦MON-POW⟧ consists of all two different monoid structures on *the same set*, plus an interpretation of _^_ respecting the "power" equations.

Notice that in the sharing approach the elements and the powers are taken from the same set. In order to avoid this situation, even in the sharing approach, there are several solutions:

1. "import" TRIV as a parameter of MON rather than normal import (i.e., using MON* of Example 26), or

2. flatten MON by directly specifying the content of the import TRIV (i.e., the sort Elt).

In the case of 1. the corresponding diagram would be

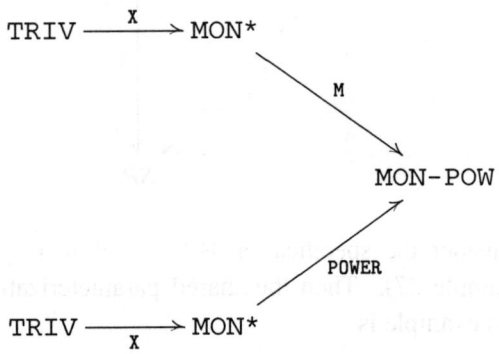

while in the case of 2. it would be

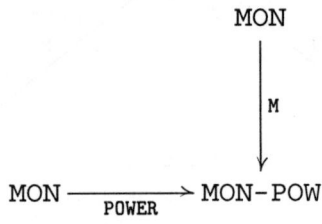

□

Shared parameter instantiation

In the sharing approach, parameter instantiation by any view might produce undesirable side-effects. For example, if the parameter shares a sub-module with the main body of the parameterized module, instantiating the parameter might change the body. Similarly, in the case of multiple parameters, the instantiation of only one of them might change the other parameter too, therefore such partial instantiation might fail to still be a parameterized module. In order to avoid such side effects we restrict the use of views in instantiations as follows:

> In the shared approach, the instantiation of parameters is restricted only to the views which preserve all the imports of the parameters, i.e., which are identities on the imports of the parameters. Also, the result of the instantiation is always an import.[4]

We know that given a parameterized module $P \xrightarrow{X} SP$, the result of instantiating by a view $v \colon P \to P'$ is an injection $X(v) \colon P' \to SP(v)$. However in the sharing approach we need to take into account the sharing between the actual value of the parameter (P') and the body of the parameterized module. Therefore, the instantiation is a co-limit of the following diagram rather than just a pushout:

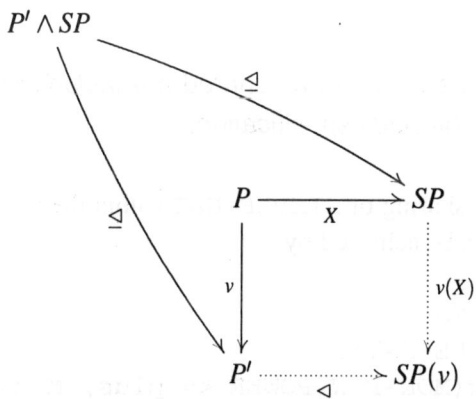

Notice that the result of the instantiation, $X(v)$, is chosen to be an import, i.e., from all isomorphic co-limits we chose the one for which $X(v)$ is an actual *inclusion* rather than an injection.

---

[4]We can relax this rule by requiring that the views used in the case of multiple parameter instantiations agree on the common imports of the parameters but are still identities on the imports common which appear in the body.

Example 35 Consider the parameterized module STRG#* of Example 30. In the shared approach, the result of the instantiation of X by the view bare-nat is an import BARE-NAT ⊴ STRG#* (bare-nat), hence BARE-NAT is shared and therefore appears only once (i.e., without any other copy of it) in STRG#* (bare-nat). □

Example 36 The instantiations of Example 32 are illegal in the sharing approach since the views involved do not preserve the import of TRIV. For doing such instantiations one should either use the parameterized monoids MON* instead of MON, or use a basic specification of monoids (without imports). □

Enforcing sharing in the non-sharing approach

In the non-sharing approach, it is often useful to have a flexible simple system for sharing (parts of) some parameters rather than enforcing the sharing by user defined collapsing of different copies of the same module. In CafeOBJ this can be achieved by the command share which has the effect of enforcing that

> The modules declared as shared are *included* rather than injected in the body specification.

Example 37 The sharing of SIMPLE-NAT within the module NAT-TIMES of Example 32 can be achieved by

```
mod* NAT-TIMES {
  share(SIMPLE-NAT)
  protecting(MON-POW(POWER <= plus, M <= plus)
    * { op _^_ -> _*_ })

  eq m:Nat * s(0) = m .
}
```

The (commutative) diagram corresponding to NAT-TIMES and illustrating the sharing of SIMPLE-NAT can be visualized as:

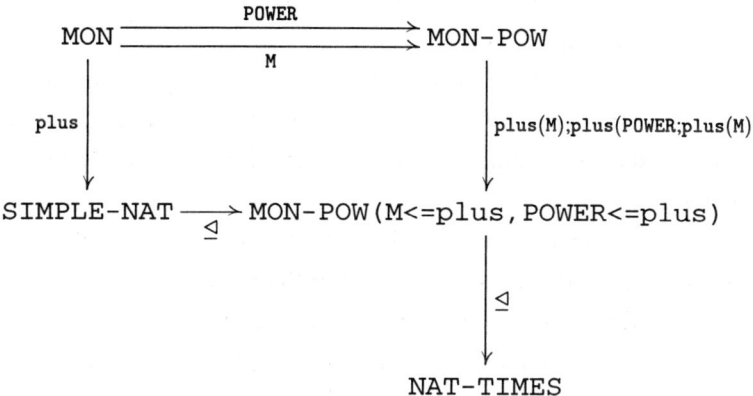

Notice that this diagram comes as a particular case of the generic diagram on instantiating multiple parameters. □

The sharing approach can be regarded as the ultimate enforcing of sharing in a non-sharing framework, in which each module is shared by default.

10 Module Expressions

Module expressions are formed as iterations of the following basic constructs.

- imports (which can be multiple),
- parameterized modules,
- translations,
- sum, and
- views and parameter instantiation.

10.1 Translations

A **module translation** implements the concept of sentence translation in a direct way. Given a specification *SP*, a translation from *SP* is just a *surjective* signature morphism $\phi: sign(SP) \to sign'$. The translation of *SP* along ϕ has, of course, *sign'* as signature and the translation $\phi(SP)$ of the sentences in *SP* as sentences. Therefore, a translation ϕ gives rise to a surjective specification morphism $\phi: SP \to (sign', \phi(SP))$. We may simplify the notation $(sign', \phi(SP))$ to just $\phi(SP)$.

The denotation $[\![\phi(SP)]\!]$ consists of all *protecting* expansions along ϕ of the models in $[\![SP]\!]$. Notice that we do not require priori that the models in $[\![\phi(SP)]\!]$ to satisfy the specification *SP* since this holds anyway by the Satisfaction Condition.

Translation of modules are useful in practice for increasing the readability of **CafeOBJ** specifications, and for controlling (either avoiding or enforcing) module sharing. They are typically used after instantiating parameters in modules. In order to avoid module sharing one has to use a "renaming" of the module items (this corresponds to a bijective translation), and for enforcing module sharing (in the non-sharing approach) one may use a collapsing translation which identifies corresponding items from two different copies of the same module.

Example 38 Consider the specification of semi-rings (without units) by reusing the specification of commutative monoids with powers (Example 28) in the non-sharing approach. This is achieved by enforcing the sharing of the two monoids.

```
mod* SRNG {
   protecting(CMON-POW *
    { sort Elt.POWER -> Srng,
      sort Elt.M      -> Srng,
      op (_;_).POWER -> _+_,
      op (_;_).M      -> _+_,
      op nil.POWER -> 0,
      op nil.M      -> 0,
      op _^_ -> _*_ })
}
```

or in more compact notation

```
mod* SRNG {
   protecting(CMON-POW *
      { sort Elt -> Srng, op _;_ -> _+_, op nil -> 0,
```

```
    op _^_ -> _*_ })
}
```

This gets a non-parameterized specification of semirings. We can get an equivalent specification parameterized by the additive monoid:

```
mod* SRNG* (M+ :: CMON) {
    protecting(CMON-POW *
        {param POWER -> M+, param M -> M+} *
            {sort Elt -> Srng, op _;_ -> _+_, op nil -> 0,
                                op _^_ -> _*_ })
}
```

by mapping both parameters of CMON-POW to the same parameter M+. □

10.2 Evaluations

Imports, sums, parameterized modules, instantiations, and translations are all instances of module expressions. As mentioned above these can be combined into complex module expressions. The evaluation of a CafeOBJ module expression is an iterative process over the structure of the module expression, which computes at the end a basic specification as the flattening of the module expression to a basic specification. Notice that the denotation of the value of the module expression can be very different from the denotation of the module expression since the process of flattening looses much information.

This reveals another important rôle played by module expressions besides the well celebrated rôle of reusing already written specifications. Module expressions permit formal description of classes of models (implementations) which cannot be described as denotations of basic specifications.

This Report promotes the systematic use of diagrams (of specification morphisms) for visualizing module expressions. In this part we showed how each atomic structuring concept can be visualized as such a diagram. The process of evaluating a module expression can be regarded as computing the co-limit of such diagrams (very much in the tradition of Clear [5] and OBJ [40]). Notice that the diagrams corresponding to module expressions do not have multiple edges between two given nodes.

Example 39 Recall the process of obtaining the multiplication of naturals as the power of the addition (Example 32). Notice the use of the translation collapsing the two copies of the naturals, thus transforming the _^_ from heterogeneous operation to a homogeneous one, and finally renaming it to _*_; we denote this translation by ϕ_1.

We can extend this process to defining the ordinary power operation on naturals as follows:

```
view nat-times from MON to NAT-TIMES {
   sort Elt -> Nat,
   op 0 -> s(0),
   op _;_ -> _*_
}

mod* NAT-POW {
   protecting(MON-POW-NAT(M <= nat-times) *
      { sort Nat.M -> Nat, sort Nat.POWER -> Nat })
}
```

Notice we again use a translation (denoted by ϕ_2) collapsing two copies of the naturals.

The process of building NAT-POW can be visualized by the following commuting diagram:

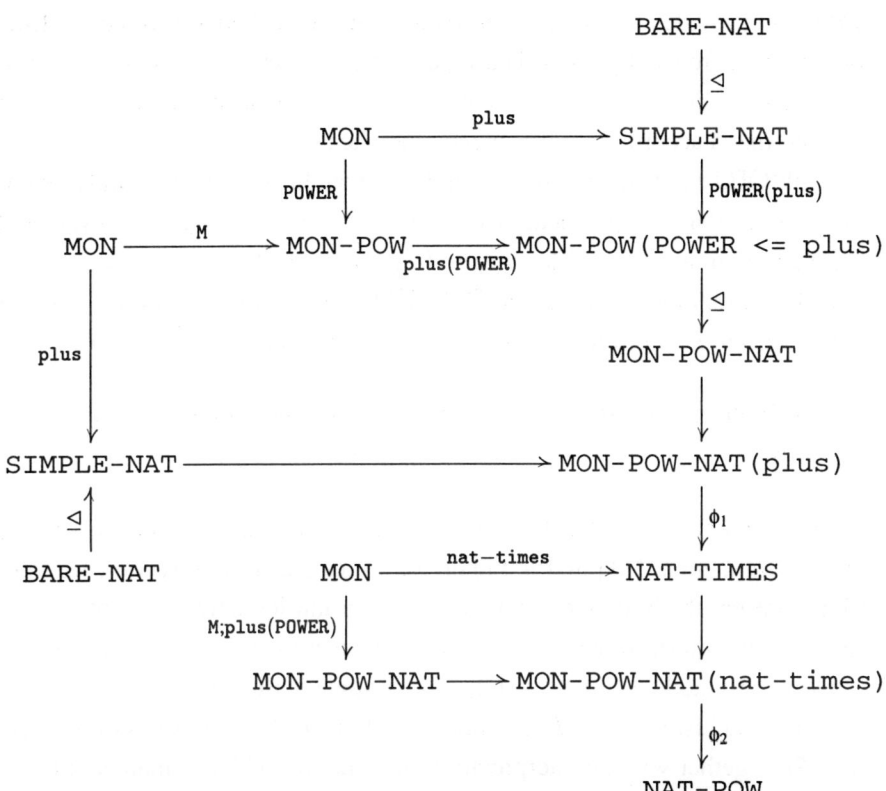

In the sharing approach the use of the views `plus` and `nat-times` are illegal since they instantiate the sort `Elt` of the import `TRIV`. In order to make them legal we might use the parameterized monoids `MON*` (Example 26) or a basic specification of `MON` (avoiding imports). Also, in the sharing approach we do not need any of the collapsing translations ϕ_1 and ϕ_2 to enforce the sharing of the naturals since this sharing is achieved by default, all we need from ϕ_1 is only the renaming of `_^_`. We can obtain the same effect in the non-sharing approach by using the command `share`:

`share(SIMPLE-NAT)`

in both `NAT-TIMES` and `MON-POW-NAT`. □

11 Built-in Modules

CafeOBJ provides a library of **built-in modules** called **standard prelude**
for the most frequently used data types, such as numbers, tuples, booleans,
identifiers, strings, etc. This list is of course open, so the standard prelude can
be customized according to the intended applications.

CafeOBJ built-in modules should be thought as models (implementations) rather than as specifications since their operations are often evaluated
as functions in a low level language. In most cases these models can be regarded as initial models for some CafeOBJ specification, however this is not
always the case, a notorious example being the real numbers.[5] Therefore:

> A built-in module is a signature together with one of its mod-
> els.

The semantics of CafeOBJ specifications with built-in modules is based
on *constraint logic* [11] and is rather sophisticated. In this section we very
briefly present the basic ideas behind the constraint logic treatment of specifications with built-in modules. Consider a built-in module whose signature is
Σ, and whose model is a Σ-model A. Then a **specification** *SP* **over** $\langle \Sigma, A \rangle$ consists of an inclusion $\Sigma \hookrightarrow SP$. A **model of** *SP* **over** $\langle \Sigma, A \rangle$ consists of a model
M of *SP* together with an interpretation of A into M, which is mathematically
modeled as a model morphism $m\colon A \to M\!\restriction_\Sigma$.

Example 40 Consider the example of a specification of the Euclidean plane
as a vector space over the real numbers.

```
mod! R2 {
  protecting(FLOAT * { sort Float -> Real })

  [ Vect ]

  op 0 : -> Vect
```

---

[5]There is no finite algebraic specification for the real numbers due to the cardinality argument. But even for the floating point representation of real numbers it would be very unpractical
to consider a specification which has this as its initial model.

```
op <_,_> : Real Real -> Vect
op _+_ : Vect Vect -> Vect
op -_ : Vect -> Vect
op _*_ : Real Vect -> Vect

vars a b a' b' k : Real

eq 0 = < 0 , 0 > .
eq < a , b > + < a' , b' > = < a + a' , b + b' > .
eq k * < a , b > = < k * a , k * b > .
eq - < a , b > = < - a , - b > .
}
```

The signature Σ of built-in sorts and operation symbols contains one sort Real[6] for the real numbers together with the usual ring operation symbols and a relation symbol $\_<\_$. The built-in model is just the usual ring of real numbers (denoted as $\mathbb{R}$) with $\_<\_$ interpreted as the usual 'strictly less than' predicate. The signature *sign*(R2) introduces a new operation symbol $<\_, \_>$ for representing the points of the Euclidean plane as tuples of real numbers, and overloads the ring operations by organizing the Euclidean plane as a vector space over the real numbers. The axioms express the basic fact that the evaluation of the ring operations on vectors is done component-wise.

A standard model for this specification, denoted $\mathbb{R}^2$, is given by the Cartesian representation of the points of the Euclidean plane, i.e., any point is represented as the tuple of its coordinates. Another model for R2 interprets the sort Vect as the set of real numbers, the $\mathbb{R}$-module[7] operations on Vect as ordinary operations on numbers, but $<\_, \_>$ as addition of numbers. We denote this model by $\mathbb{R}+$.

For both $\mathbb{R}^2$ and $\mathbb{R}+$, the interpretation of the built-in reals is just the identity morphism on $\mathbb{R}$ which is the reduct of both $\mathbb{R}^2$ and $\mathbb{R}+$ to Σ. The denotation $[\![\text{R2}]\!]$ consists only of the initial model (but in the more general sense of constraint logic [11]), which is $\mathbb{R}^2$. □

The concept of **constraint sentence** and **constraint satisfaction** for spec-

---

[6]Obtained here by renaming the sort Float of the imported built-in CafeOBJ module FLOAT implementing the real numbers as floating point reals.

[7]In the sense of linear algebra.

ifications with built-in modules generalize the usual CafeOBJ sentences and satisfaction in the style of constraint logic [11] by allowing the elements of the built-in modules to take part in term formation. Such terms are called **generalized polynomials** [11]. Constraint sentences are thus formed by generalized polynomials and the concept of constraint satisfaction uses the interpretation morphism from the built-in model for evaluating the generalized polynomials.

Example 41 In the context of the above example

```
eq < 3.14 * X , Y > + - < Y , 3.14 * X > = 0 .
```

is a *constraint equation* which is satisfied by $\mathbb{R}+$ but *not* by $\mathbb{R}^2$. □

Specifications with built-in modules admit loose and tight denotations in the style of ordinary specifications, the construction of initial models, and more generally, free expansions, carrying over from the ordinary specifications. Also, the completeness of proof calculi extends to specifications with built-in modules [11]. We do not insist here on these constructions which in the case of specifications with built-in modules are significantly more sophisticated than in the ordinary case, but we rather suggest the reader consult [8] or [11] which constitute the basic references for constraint logic. However, at the abstract level of category-based equational logic [9, 8, 32], constraint logic is just one of its instances [11], hence its equational essence.

III Proof Technologies

12 Rewriting

Rewriting is the basis of the CafeOBJ operational semantics and constitutes the full operational semantics for the equational and rewriting specification by regarding the equational/rewriting specifications as **term rewriting systems** (abbreviated **TRS**).

Given a signature $(S, \leq, \Sigma)$, a **TRS-rule** is given by

$$(\forall X)\, t \ \texttt{->}\ t' \ \textbf{if}\ C$$

where t, t', C are Σ-terms such that $var(C) \subseteq X = var(t) \supseteq var(t')$ and C: Bool. Although apparently they are the same, we distinguish between the CafeOBJ transitions (Section 3.3) which play a denotational rôle of sentences in CafeOBJ rewrite specifications, and TRS-rules which play a purely operational rôle.

A TRS is a finite collection of TRS-rules. Given a fixed TRS, then a Σ-term t_0 **rewrites** (in one step) to the Σ-term t_1 iff there is a TRS-rule

$$(\forall X)\, t \ \texttt{->}\ t' \ \textbf{if}\ C$$

such that $\theta(C) = \texttt{true}$ and $t_0 = c(\theta(t))$ and $t_1 = c(\theta(t'))$ for some context (called **rewrite context**) c and some substitution θ. This is denoted as $t_0 \to t_1$. The transitive-reflexive closure of $\to$ is denoted as $\xrightarrow{*}$. A TRS is **(ground) confluent** iff for any (ground) term t_0, if $t_0 \xrightarrow{*} t_1$ and $t_0 \xrightarrow{*} t_2$, then there exists t_3 such that $t_1 \xrightarrow{*} t_3$ and $t_2 \xrightarrow{*} t_3$,

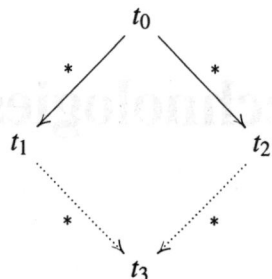

and it is **(ground) terminating** iff there are only finite rewrite chains from any (ground) term t. A term t is in **normal form** iff there is no rewrite from t. When the TRS is confluent and terminating, then each term t has a unique normal form $\mathrm{nf}(t)$ such that $t \xrightarrow{*} \mathrm{nf}(t)$.

Rewriting can be generalized to congruence classes of terms. Given a fixed TRS and a set E of equations, then a congruence class $t_0/ \equiv^E$ rewrites (in one step) to $t_1/ \equiv^E$ iff there exists $t_0' \to t_1'$ with $t_0 \equiv^E t_0'$ and $t_1 \equiv^E t_1'$. All other rewriting concepts lift easily from ordinary rewriting to rewriting modulo congruence classes.

A good and almost complete exposition of rewriting can be found in [28].

12.1 Equational specification

Each equational specification can be regarded as a TRS by orienting the equations as TRS-rules. While this can be achieved for all equations, some of the resulting TRS-rules would generate non-termination of rewriting; associativity and commutativity being typical examples. We call such equations **non-orientable**, and we treat them as equations for rewriting modulo.

When the resulting TRS is confluent, we have the completeness of rewriting:

$$SP \models (\forall\emptyset)\ t\ =\ t' \text{ iff there exists } t'' \text{ such that } t \xrightarrow{*} t'' \text{ and } t' \xrightarrow{*} t''.$$

If the TRS is also terminating, then the last condition can be simplified as

$$SP \models (\forall\emptyset)\ t\ =\ t' \text{ iff } \mathrm{nf}(t) = \mathrm{nf}(t')$$

The CafeOBJ command "reduce" reduces a given ground term to its normal form by using the TRS formed *only by the equations* of the specification (thus ignoring the transitions). Although both visible and hidden sorted terms can be given as input to reduce, this command is rarely used for evaluations of hidden sorted terms because the meaning of reducing a term by reduce is that of semantic equality.

Because behavioural equations denote behavioural equivalences rather than equality, their use in the reduction process is rather subtle. The following principle governs the (safe) use of behavioural equations in reduction:

> For each rewriting step in which a behavioural equation is used, the corresponding (rewriting) context has a sub-context[1] which is *behaviourally coherent.*

Behaviourally coherent contexts are contexts of visible sort without any operation on hidden sorts which is neither behavioural nor behaviourally coherent on top of the variable. The following figure might help visualising this condition:

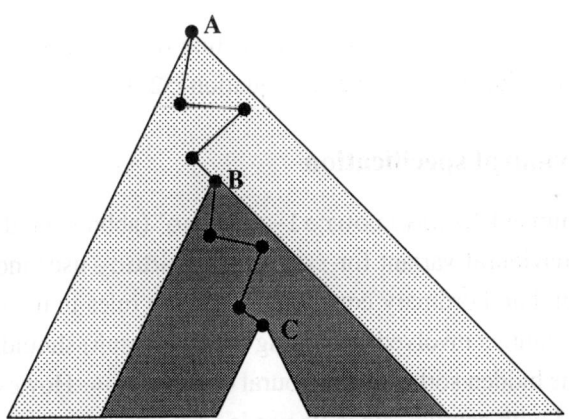

Here A is the top position of the term to be reduced (represented by the big triangle), and C is the position of the sub-term (represented by the white triangle) to which the rule is applied. The rewriting context is represented by

---

[1]A **sub-context** is a sub-term of the context which is a context itself, i.e., contains the variable of the context.

the whole gray area, and the behaviourally coherent sub-context by the dark gray area (with the top at B). The condition says that the sort of the operation at position B is visible, and that on the path between B and C there are no operations which are non-behavioural and not coherent.

Example 42 In the context of the "history sensitive storage" Example 14,

```
red rest(put(e, st)) == st .
```

gives `false`, but

```
red put(get(rest(put(e1, put(e2, st1)))), st2) ==
    put(e2, st2) .
```

gives `true` by rewriting the left-hand side by the behavioural equation (the behavioural coherent sub-context being the subterm at `get`) to `put(get(put(e2, st1)), st2)`, and then by the other equation.

Another example when real equality between hidden sorted terms is inferred with `reduce` is given by HSS-NAT-PROOF. Then

```
red put(1 + 2, rest*(st, 0)) == put(3, st) .
```

gives `true` after the natural numbers reduction of 1 + 2 to 3, and after the reduction on the hidden sort using the equation p2. □

12.2   Behavioural specification

CafeOBJ command "`behavioural-reduce`" (abbreviatted as `breduce`) provides a behavioural variant for `reduce`. `breduce` uses indiscriminately both behavioural and ordinary equations as rewrite rules in the reduction process, but all equalities involved (including the final result) should always be interpreted on the hidden sorts as behavioural equality `=b=`. However, `breduce` should be used with caution, since its use is sound only if all non-behavioural operations are *behaviourally coherent*.

Example 43 In the context of the "history sensitive storage" Example 14,

```
bred rest(put(e, st)) .
```

gives `st`,

```
    bred rest(put(e, st)) =b= st .
```

gives true,

```
    bred rest(put(e, st)) == st .
```

is treated as the former, and

```
    red rest(put(e, st)) =b= st .
```

gives true because =b= is evaluated with breduce. ☐

Behavioural coherence of operations can be checked with the system by simple proof scores as in the following example:

Example 44 In the latter BAG specification of Example 15, the behavioural equivalence is =*=, so the coherence of put and get can be proved as follows:

```
    ops b1 b2 :   -> Bag .
    ops e  e' :   -> Elt .
    var E : Elt
-- hypothesis
    eq get(b1, E) = get(b2, E) .
-- case e =/= e'
red get(put(e, b1), e') == get(put(e, b2), e') .
red get(take(e, b1), e') == get(take(e, b2), e') .
-- case e == e'
red get(put(e, b1), e) == get(put(e, b2), e) .
red get(take(e, b1), e) == get(take(e, b2), e) .
```

Notice that for correctly running this proof score, the coherent attribute should be (temporarily) commented out in the specification. ☐

12.3   Rewrite specification

In the case of rewrite specification, the corresponding TRS consists of both orientable equations and transitions. In contrast with the equational specification case, here confluence does not play any significant rôle. Because in the case of rewrite specification the operational semantics is encoded directly at the level of denotational semantics, the completeness of rewrite specification operational semantics is just the completeness of the rewriting logic proof theory, i.e.,

$$t \xrightarrow{*} t' \quad \text{iff} \quad SP \models (\forall \emptyset)\ t \Rightarrow t'$$

The CafeOBJ command "execute" reduces a given ground term to a normal form by using the TRS formed by both equations and transitions modulo the non-orientable equations. execute accepts only visible sorted terms.

13   Induction

An **inductive property** is a sentence satisfied by the initial model of a specification. By the basic completeness results for equational and rewriting logic, any (logical) consequence of a specification is also an inductive property, however the opposite is not true.

Example 45   Consider the specification SIMPLE-NAT of additive naturals of Example 23 but without the commutativity of _+_. The initial model of this is still **Nat+** expanding (*protecting*) the initial model **Nat** of BARE-NAT with an interpretation of _+_ as addition of naturals. In this case the commutativity of _+_ is an inductive property since

Nat+ $\models (\forall \text{M}, \text{N}:\text{Nat})\ \text{M} + \text{N} = \text{N} + \text{M}$

However, there exists models (of the specification SIMPLE-NAT without the commutativity of _+_) *not satisfying* this equation. For example, consider the *protecting* expansion **Strg+** of the model **Strg** (Example 7) with the interpretation of _+_ as string concatenation. In **Strg+** the commutativity of addition fails to be true, since, for example, *a.b* is different from *b.a*. □

It is important to notice that unlike ordinary properties, the inductive properties are *not* preserved by module imports.[2] However, inductive properties are preserved by the *protecting* imports, but may fail to be preserved in the presence of weaker importation modes.

---

[2]In other words, inductive satisfaction cannot be an institution satisfaction relation.

Example 46 Within the framework of Example 23, assuming SIMPLE-NAT does not specify _+_ as commutative, the commutativity of _+_ is not preserved as inductive property along the import SIMPLE-NAT ⊴ NAT-OMEGA+. In this case the equation

```
eq N:Nat + omega = omega .
```

does not hold in the initial model of NAT-OMEGA+. □

Example 47 In Example 23, the equation

```
eq omega + N:Nat = omega .
```

is an inductive property of the following:

```
mod! NAT-OMEGA++ {
   protecting(SIMPLE-NAT + NAT-OMEGA)

   eq omega + omega = omega .
}
```

and can be proved by the following simple proof score:

```
-- n = 0
red omega + 0 == omega .
-- inductive hypothesis
   op n :   -> Nat .
   eq omega + n = omega .
red omega + s(n) == omega .
-- on infinity
red omega + omega == omega .
```

which is just an induction proof (with case analysis) on the structure of the initial model of NAT-OMEGA++. This very common proof technique is called **structural induction.** . □

In general, inductive properties cannot be proved by simple equational (or rewriting logic) reasoning since they can go beyond ordinary consequences of the specifications. There are several induction schemes and much work has been done for developing advanced tools for supporting induction proofs, but these are beyond the scope of CafeOBJ (they rather belong to the environment around CafeOBJ). For an elegant and highly conceptual discussion on induction schemes we recommend [28].

14 Coinduction

Coinduction is a proof technique for proving behavioural properties (i.e., properties of behavioural specifications). In principle one may successfully use directly the context-based definition of behavioural satisfaction (see Section 4.3). This amounts to equational (or rewriting) based proofs organized by induction on the structure of the behavioural contexts. However, in practice this proof method easily leads to very complex proofs, which in the case of specifications of real software systems may prove unmanageable [42]. Coinduction appears as an efficient alternative to context induction proofs, however it requires more understanding of the nature of the specified system.

14.1 Proving behavioural equivalence

Proving behavioural equations is the same as proving the behavioural equivalence of the states denoted by the sides of the equation. Then a coinduction proof for a behavioural equation $(\forall X)\, t = t'$ consists of the following steps:

I. define a hidden equivalence $\_R\_$,

II. prove that $\_R\_$ is a congruence,

III. prove that $t\ R\ t'$.

The correctness of the coinduction method follows from the fundamental result characterizing the behavioural equivalence as the largest hidden congruence (see Section 4.3). Notice that $\_R\_$ does not refer to a particular relation on a particular model but it is rather interpreted in all models as any other operation symbol. Finally, it is easy to notice that $\_R\_$, as Bool-valued operation, is always behaviourally coherent.

Example 48 Consider Example 14 of "history sensitive storage". Here we define a coinduction relation for "history sensitive storages", prove it is a hidden congruence, and finally we illustrate how to use it to prove some behavioural properties.

The coinduction relation $\_R\_$ is defined mathematically as an infinitary conjunction indexed by the natural numbers:

$$s1\ R\ s2 \text{ iff } \forall n \in \omega\ \texttt{get}(\texttt{rest}*(s1,n)) = \texttt{get}(\texttt{rest}*(s2,n))$$

This definition can be expressed directly in CafeOBJ by using the power of universal quantification for CafeOBJ equations by parameterizing _R_ as follows:

```
mod COINDUCTION-REL {
  protecting(HSS-NAT-PROOF)

  op _R[_]_ : Hss Nat Hss -> Bool

  vars S1 S2 : Hss
  var N : Nat

  eq [ R ] : S1 R[N] S2 =
             get rest*(S1, N) == get rest*(S2, N) .
}
```

In order to prove that _R_ is a congruence, we have to assume that s1 R s2 and show this is preserved when applying all "methods" and "attributes".

```
open HSS-NAT-PROOF .
  ops m n : -> Nat .
  ops e e1 e2 : -> Elt .
  ops st s1 s2 : -> Hss .

  eq [ hyp ] : get rest*(s1, N) = get rest*(s2, N) .
```

We prove it first for get,

```
  start get s1 == get s2 .
  apply -.p2 within term .
  apply .hyp within term .
  apply reduce at term . -- it should be true
```

then for put by case analysis on n,

```
  red put(e, s1) R[s(n)] put(e, s2) .
  red put(e, s1) R[0] put(e, s2) .
```

and finally for rest.

```
start (rest s1) R[n] (rest s2) .
apply .R within term .
apply -.p1 within term .
apply .hyp within term .
apply reduce at term .
```

This completes the proof that R is a hidden congruence. In this proof score, we use the special CafeOBJ command apply which enables the user to control the reduction process and use the equations in the specification from right to left. apply uses the labels of the equations, no sign the equation is applied in the usual way (i.e., from left to right), and – means the equation is applied opposite to the usual way (i.e., from right to left). apply reduce means reduction to normal form.

Now, for proving the behavioural equivalence of t and t' by coinduction, it is enough to prove that t R t'.

The following is the proof score of
HSS-NAT $\models$ rest rest put$(e1,($put$(e2,st)))=st$:

```
red (rest rest put(e1, (put(e2, s))))) R[n] s .
```

and

```
red (rest*(put(e, s), s(m))) R[n] rest*(s, m) .
```

is the proof score of HSS-NAT $\models$ rest*$($put$(e,st),s(n))=$rest*(st,n).
□

Other examples of coinduction proofs for behavioural equivalence can be seen in the Methodologies Part, where we also show how coinduction relations can be reused when composing objects.

14.2 Proving behavioural transitions

In CafeOBJ behavioural transition properties are expressed as behavioural transitions . In the case of behavioural transitions we have less flexibility, in the sense that the coinduction hidden equivalence is always _=*=_. Then a coinduction proof for a behavioural transition $(\forall X)\ t \Rightarrow t'$ consists of the following steps:

I. prove that _=*=_ is a congruence,

II. prove that $attr(t) ==> attr(t')$ for all (parameterized by the values of the data) attributes $attr$.

The correctness of coinduction for behavioural transitions is proved in [10].

Example 49 Consider the example of the nondeterministic counter COUNTER-NNAT of Example 33. In this case _=*=_ is a congruence so we have to do only the step II. Therefore, consider the following behavioural transition property:

```
btrans add(M | N, C) => add(M, C) .
```

Then this gets very simple proof score:

```
open COUNTER-NNAT .
  ops m n : -> Nat .
  op counter : -> Counter .
red read add(m | n, counter) ==> read add(m, counter) .
```

□

IV Methodologies

This part is devoted to the presentation of several important CafeOBJ methodologies emphasizing the use of new language features, such as behavioural concurrent specification and rewriting logic. We therefore deliberately neglect the methodologies related to the traditional equational specification part of the language, since this has been already treated in the vast algebraic specification literature. This topic is by no means closed, new methodologies will certainly be proposed in the future; we present here only a few of them which were developed and well understood at the time this Report was completed.

15 Nondeterminism

Nondeterminism can be handled in CafeOBJ in several different ways corresponding to the several different paradigms implemented by CafeOBJ, such as rewriting logic and behavioural specification. In this section we illustrate comparatively two different ways to treat nondeterminism by means of a simple example: the nondeterministic choice of natural numbers.

15.1 In Rewriting Logic

This is a classical example used in the literature [49] to illustrate the semantics and the power of rewriting logic.

```
mod! NNAT-RWL {
  extending(NAT)

  op _|_ : Nat Nat -> Nat
```

```
vars M N : Nat

trans N | M => N .
trans N | M => M .
}
```

In this example we used the built-in module NAT of the natural numbers. The nondeterministic choice is modeled via the operation _|_ and the two transitions.

Notice that the sort of natural numbers gets "nondeterministic naturals" as new elements, hence *extending*. Transitions are responsible for the nondeterministic choice. Also, all operations inherited from NAT are extended automatically on the nondeterministic naturals. For example, a property such as

$$3 \leq (4|4|5)$$

means that whatever choice we make from (4|4|5) the result is not going to be less than 3. The CafeOBJ proof score for this property is as follows:

```
open NNAT .
red (3 <= ( 4 | 4 | 5 )) ==> false .
```

The result of running this proof score is false, which means that there is no possible transition from 3 <= (4 | 4 | 5) to false. Ultimately, this means that 3 <= (4 | 4 | 5) *for all possible transitions*, which is exactly what we want.

15.2   In Behavioural Specification

The HSA version of this problem has a very simple specification too:

```
mod* NNAT-HSA {
  protecting(NAT)

  *[ NNat ]*
```

```
op [_]  : Nat -> NNat
op _|_  : NNat NNat -> NNat {coherent}
bop _->_  : NNat Nat -> Bool

vars S1 S2 : NNat
vars M N : Nat

eq [M] -> N = M == N .
eq (S1 | S2) -> N = (S1 -> N) or (S2 -> N) .
}
```

Here the space of nondeterministic naturals is hidden (modeled by NNat), the nondeterministic choice is modeled via the non-behavioural operation _|_ and via the "attribute" _->_ which is formulated as a predicate expressing the actual choice. Notice that the behavioural equivalence is _=*=_ and that _|_ is *behaviourally coherent.*

   In this setup, associativity, commutativity, or idempotence (which are natural expected properties of nondeterministic choice) appear as behavioural properties and they have very simple proof scores. Take commutativity, for example:

```
open NNAT-HSA .
  ops s1 s2 :  -> NNat .
  op n :   -> Nat .
red (s1 | s2) -> n == (s2 | s1) -> n .
```

A property such as $3 \leq (4|4|5)$ gets again very simple proof, but at the cost of extension by hand of the predicate $\leq$ as an "attribute" on NNat:

```
mod* NNAT-HSA<= {
  protecting(NNAT-HSA)

  bop _<=_ : Nat NNat -> Bool

  vars M N : Nat
  vars S1 S2 : NNat
```

```
    eq  N <= [M] = N <= M .
    eq N <= (S1 | S2) = (N <= S1) and (N <= S2) .
}
```

Then the proof is as follows:

```
open NNAT-HSA<= .
red 3 <= ([4] | [4] | [5]) .
```

Notice also the complexity of such computation is $O(n)$ while the complexity of the corresponding RWL computation is $O(n!)$ since behavioural abstraction allows cutting dramatically the transition
space. On the other hand, the relative disadvantage of having to extend by hand the built-in predicates (operations) from Nat to NNat can be overcome by the use of parameterization since the equations corresponding to these extensions are essentially the same.

16 Concurrent Object Composition

In this section we present the basic methodology for composing concurrently objects in CafeOBJ within the framework of behavioural specification. The CafeOBJ methodology for composing objects not only provides support for reusability of source code, but also supports reusability of proofs, which is of crucial importance for the verification stage. Also, concurrent object composition gets more reusability power than the alternative (more traditional) *inheritance* method; for more details on this topic the reader may wish to consult [44]. Notice that in all examples in this Report, each object is specified by one module. This is just a *recommended* specification style and not a strict conceptual correspondence between objects and modules.

Concurrent object composition in CafeOBJ can be visualized in the following figure which uses the OMT [17] like notation to represent relations between objects.

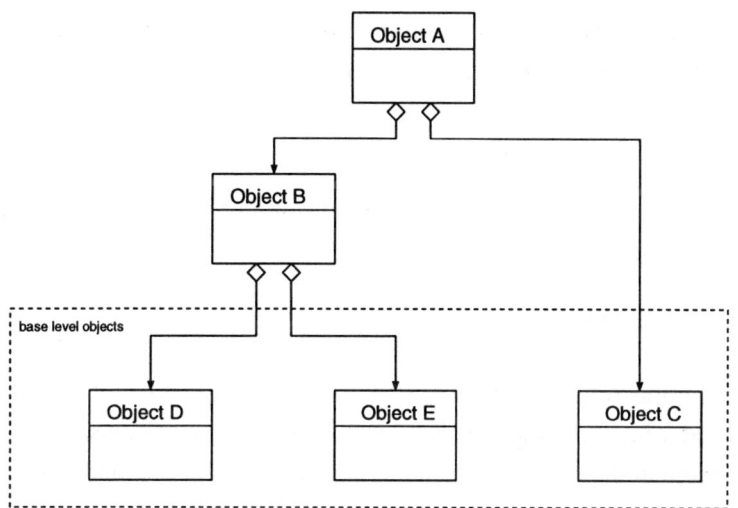

In the OMT notation, each object is represented by a box containing the object name (which is the CafeOBJ case is given by the name of the hidden sort representing its states) followed by the list of its attributes and by the list of its methods. The arrows between objects represent relationships between a composed object and its components (in CafeOBJ these are projection operations).

16.1 Specifying Composed Objects

Projection operations play the essential rôle in object composition. They get the states of the composing objects from the states of the composed objects and all methods of the composed object are related to the methods of the composing objects using these projection operators.

An operation $\pi_n : h \to h_n$ is a **projection operation** if:

- h is a hidden sort of the composed object O,
- h_n are hidden sorts of the composing objects O_n,
- for each attribute a of O, there exists a composing object O_n, and a visible O_n-context c_n such that:

$$a = \pi_n; c_n,$$

- for each method $m : h \to h$ of O, for all composing object O_n, there exists a sequence of methods m_n such that:

$$m; \pi_n = \pi_n; m_n,$$

- for each constant $c :\to h$ of O, for all composing objects O_n, there exists a constant $c_n :\to h_n$ such that:

$$\pi_n(c) = c_n$$

This definition refers to static systems (i.e., configuration of the system is unchanged when running), see Section 17 for dynamic systems.

Notice that the equalities defining the attributes and methods for the composed objects are strict (i.e., not behavioural) equations and that projection operators can be either ordinary or behavioural. Using non-behavioural projections has the advantage to enable a user controlled selection of the attributes on the composed object, but they have the disadvantage of possibly restricting the computations involving behavioural equations. However, when the projections are *behaviourally coherent* we have the same computational power as in the case of behavioural projections.

As shown in the above figure, the structure of such a composition is a DAG (directed acyclic graph). A **base level object** is an object without projection operators. Two methods of a composed object are in the same **methods group** when they are related to the same composing object. If the method in a composed object relates to several method in different composing objects then there is an overlapping among the method groups.

Object composition can be fully concurrent or synchronized. In the case of synchronized concurrent composition, the concurrency between composing objects is partial. Synchronization happens when:

1. the projected state of the composed object (via a projection operation) depends on the state of a different (from the object corresponding to the projection operation) composing object, and

2. methods of the composed object change simultaneously states of several composing objects.

These conditions amount to refining the definition of projection operations by considering conditions for the equations defining the attributes and methods of the composed object in the following way:

- each condition is a finite conjunction of equalities between terms of the form $\pi_n; c_n$ (where π_n is a projection operator and c_n is an O_n-context) or terms in the data signature, and

- the conditions corresponding to a given left hand side are mutually disjoint and their disjunction is always true.

In the following we illustrate this methodology for concurrently composing objects by an example of a counter with switch, which has methods for adding or subtracting a natural number to the counter, depending on the position of the switch. There are two composing objects: COUNTER and SWITCH. We first specify SWITCH:

```
mod! ON-OFF {
  [ Value ]

  ops on off : -> Value
}

mod* SWITCH {
  protecting(ON-OFF)

  *[ Switch ]*

  op init-sw : -> Switch
  bop on_ : Switch -> Switch      -- method
  bop off_ : Switch -> Switch     -- method
  bop state_ : Switch -> Value    -- attribute

  var S : Switch

  eq state init-sw = off .
  eq state(on S) = on .
```

```
  eq state(off S) = off .
}
```

and then the COUNTER:

```
mod* COUNTER {
  protecting(INT)

  *[ Counter ]*

  op init-counter : -> Counter
  bop add : Int Counter -> Counter      -- method
  bop read_ : Counter -> Int            -- attribute

  var I : Int
  var C : Counter

  eq read init-counter = 0 .
  eq read add(I, C) = I + read C .
}
```

Finally we compose SWITCH with COUNTER as shown by the following OMT-like diagram

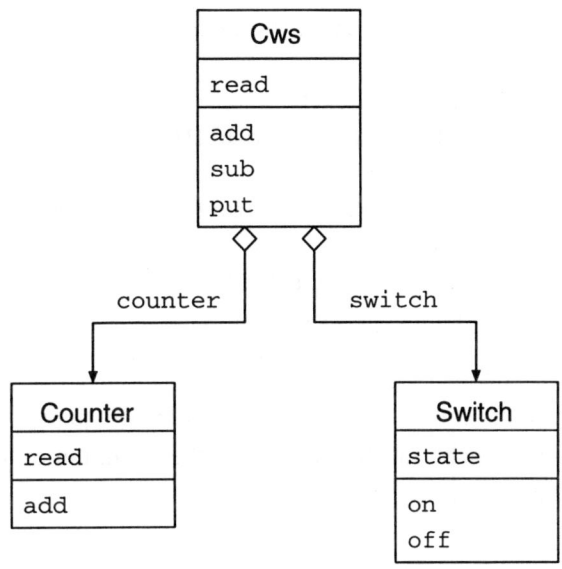

```
mod* COUNTER-WITH-SWITCH {
  protecting(SWITCH + COUNTER)

  *[ Cws ]*

  op init :  -> Cws
  bop put : Int Cws -> Cws         -- method
  bop add_ : Cws -> Cws            -- method
  bop sub_ : Cws -> Cws            -- method
  bop read_ : Cws -> Int           -- attribute
  bop counter_ : Cws -> Counter    -- projection
  bop switch_ : Cws -> Switch      -- projection

  var N : Int
  var C : Cws

  eq read C = read(counter C) .   -- abbreviation

  eq switch(init) = init .
  eq switch put(N, C) = switch C .
  eq switch add(C) = on(switch C) .
```

```
  eq switch sub(C) = off(switch C) .

  eq counter(init) = init .
  ceq counter(put(N, C)) = add(N, counter(C))
      if state(switch(C)) == on .
  ceq counter(put(N, C)) = add(-(N), counter(C))
      if state(switch(C)) == off .
  eq counter add(C) = counter C .
  eq counter sub(C) = counter C .
}
```

The synchronization can be seen in the conditional equations which corresponds to the first synchronization case (i.e., the definition of the counter depends on the state of the switch).

16.2   Verification of Composed Objects

The verification process of composed objects in CafeOBJ can reuse the behavioural equivalences of the composing objects. In many cases base level objects are rather simple and they have simple behavioural equivalence (coinduction) relations. The reusability at the verification stage relies on the following compositionality result [44] for behavioural equivalence:

Theorem 50 Given the states s and s' of a composed object, then:

$$s \sim s' \text{ if } \pi_n(s) \sim_n \pi_n(s') \text{ for all } O_n \in CObj$$

where $\sim$ is the behavioural equivalence in the composed object, $CObj$ is a set of composing objects, $\sim_n$ is the behavioural equivalence of the composing object O_n, and π_n is the projection operator to the composing object O_n. If all projection operators are behavioural then

$$s \sim s' \text{ if and only if } \pi_n(s) \sim_n \pi_n(s') \text{ for all } O_n \in CObj$$

□

In the counter with switch example, the behavioural equivalence of composing objects is just the default coinduction relation and automatically provided by the CafeOBJ system. So, from the above theorem, we can reuse the proofs of behavioural equivalence of the composing objects and get the behavioural equivalence of counter with switch.

```
open COUNTER-WITH-SWITCH .
  op _R_ : Cws Cws -> Bool    {coherent}
  vars C1 C2 : Cws

  eq C1 R C2 =  switch(C1) =*= switch(C2) and
                counter(C1) =*= counter(C2) .
```

For example, by using this, we can prove the following behavioural property:

```
-- reduce in % :
  put(m,add(put(n,sub(init-cws)))) R
  add(put(n,sub(put(m,add(init-cws))))))
true : Bool
```

Notice the crucial rôle played by the add at the top of the right hand side of the previous property, since without it the SWITCH object would be in behaviourally non-equivalent states.

16.3 Correctness of Composition

This is based on the idea that a composition is correct when the composed object is the refinement of its components and for the concurrent part the commutativity equations corresponding to the concurrency of methods/attributes belonging to different components hold. This follows some early work on concurrent composition of [31].

For example, we can show that the COUNTER-WITH-SWITCH is a correct composition of COUNTER and SWITCH as follows. In order to express properly the morphisms used in the refinement proof, we need the following "derived" method:

```
bop addc : Int Cws -> Cws    -- a derived method

var C : Cws
var N : Int
```

```
ceq addc(N, C) = put(N, C)
                if state(switch C) == on .
ceq addc(N, C) = put(-(N), C)
                if state(switch C) == off .
```

To prove that counter with switch is a correct composition of SWITCH and COUNTER, we define the following "synchronization morphism":

- ψ_1: SWITCH $\rightarrow$ COUNTER-WITH-SWITCH such that:

$$\psi_1(\texttt{init}) = \texttt{init-cws}$$
$$\psi_1(\texttt{on}) = \texttt{add}$$
$$\psi_1(\texttt{off}) = \texttt{sub}$$
$$\psi_1(\texttt{state}) = \texttt{switch; state}$$

- ψ_2: COUNTER $\rightarrow$ COUNTER-WITH-SWITCH such that:

$$\psi_2(\texttt{init}) = \texttt{init-cws}$$
$$\psi_2(\texttt{add}) = \texttt{addc}$$
$$\psi_2(\texttt{read}) = \texttt{read}$$

We prove that COUNTER-WITH-SWITCH refines SWITCH via ψ_1:

```
red state switch add(c) == on .
red state switch sub(c) == off .
```

We prove that COUNTER-WITH-SWITCH refines COUNTER via ψ_2:

```
--> case 1:
  eq state(switch c) = on .
  red read addc(i, c) == i + read c .
--> case 2:
  eq state(switch c) = off .
  red read addc(i, c) == i + read c .
```

We prove the commutativity equations corresponding to the methods:

```
--> case 1:
  eq state(switch c) = on .
  red add(addc(i, c)) R addc(i, add(c)) .
  red sub(addc(i, c)) R addc(i, sub(c)) .
--> case 2:
  eq state(switch c) = off .
  red add(addc(i, c)) R addc(i, add(c)) .
  red sub(addc(i, c)) R addc(i, sub(c)) .
```

Finally, we prove the commutativity equations corresponding to the attributes:

```
  red state(switch put(i, c)) == state(switch c) .
  red read(counter add(c)) == read(counter c) .
  red read(counter sub(c)) == read(counter c) .
```

17 Dynamic Systems of Objects

Dynamic systems are different from static systems in that the configuration of
the system changes when the system is running. The key point is that we need
some kind of identifiers to manage object creation and deletion.

The definition of **projection operations** for dynamic objects is the same
as in the static case (see Section 16) except that a projection operation for a
dynamic object has an object identifier and a composed object as its arity. An
operation $\pi_n: ID_n h \rightarrow h_n$ is a projection operation if

- h is a hidden sort of the composed object O,
- h_n are hidden sorts of the composing objects (the index n corresponding
  to the same class of objects),
- ID_n is the set of identifiers for the objects of class n, and
- the last three conditions of the definition for the static case hold under
  the "currying" of π.

Static systems appear as dynamic systems in which classes have a fixed number of objects. For example, the counter with switch example has two classes

(for the composing objects) with one object each. The following example of
a telephone system has only one class (for the composing objects) but with a
dynamic numbers of objects and we have full concurrency between the composing objects. A telephone system consists of several individual telephone
clients and it has methods for adding a new telephone into the system or for
deleting an existing one.

```
mod! NUMBER {
  protecting(NAT * { sort Nat -> Number })

  [ Number < Number* ]

  op no-tel : -> Number*    -- for errors
}

mod! UNIT {
  protecting(NAT * { sort Nat -> Unit })

  op no-tel : -> Unit    -- for errors
  eq no-tel = 0 .
}

mod* TELEPHONE {
  protecting(NUMBER + UNIT)

  *[ Tel ]*

  op no-tel : -> Tel        -- for errors
  op init-tel : Number  -> Tel    -- initial state
  bop call : Unit Tel -> Tel      -- method
  bop clear : Tel -> Tel          -- method
  bop unit : Tel -> Unit          -- attribute
  bop number : Tel -> Number*     -- attribute

  var NUM : Number
  var U : Unit
  var T : Tel
```

```
    eq unit(init-tel(NUM)) = 0 .
    eq unit(call(U, T)) = U + unit(T) .
    eq unit(no-tel) = no-tel .
    eq unit(clear(T)) = 0 .

    eq number(init-tel(NUM)) = NUM .
    eq number(call(U, T)) = number(T) .
    eq number(no-tel) = no-tel .
    eq number(clear(T)) = number(T) .
}
```

Here `call` is the method which makes a call for a certain amounts of units, `clear` is the method which resets the telephone, and `unit` is the attribute giving the total amounts of units owed at a certain moment.

```
mod* TELEPHONE-SYSTEM {
    protecting(TELEPHONE)

    *[ TelSys ]*

    op init-sys : -> TelSys              -- initial state
    bop tel : Number TelSys -> Tel              -- attribute
    bop add-tel : Number TelSys -> TelSys     -- method
    bop del-tel : Number TelSys -> TelSys     -- method
    bop call : Number Unit TelSys -> TelSys -- method
    bop pay : Number TelSys -> TelSys         -- method

    vars NUM NUM' : Number
    var U : Unit
    var TS : TelSys

    eq tel(NUM, init-sys) = no-tel .
    ceq tel(NUM, add-tel(NUM', TS)) = init-tel(NUM)
        if NUM == NUM' .
    ceq tel(NUM, add-tel(NUM', TS)) = tel(NUM, TS)
        if NUM =/= NUM' .
    ceq tel(NUM, del-tel(NUM', TS)) = no-tel
```

```
        if NUM == NUM' .
  ceq tel(NUM, del-tel(NUM', TS)) = tel(NUM, TS)
        if NUM =/= NUM' .
  ceq tel(NUM,call(NUM',U,TS)) = call(U,tel(NUM,TS))
        if NUM == NUM' .
  ceq tel(NUM, call(NUM', U, TS)) = tel(NUM, TS)
        if NUM =/= NUM' .
  ceq tel(NUM, pay(NUM', TS)) = tel(NUM, TS)
        if NUM =/= NUM' .
  ceq tel(NUM, pay(NUM', TS)) = clear(tel(NUM, TS))
        if NUM == NUM' .
}
```

Here `tel` is the indexed family of projection operations, `add-tel` and `del-tel` add and delete client telephones into the system, `call` lifts client telephone calls to the system level, and `pay` makes the payment for the amount of units owed by the client to the system.

The following OMT-like figure might help visualize this specification:

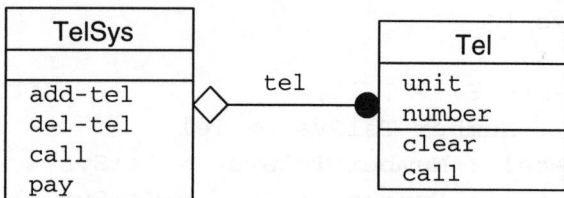

The following compositionality result [44] refines the corresponding result for static systems:

Corollary 51 Given the states s and s' of a composed object, then:

$$s \sim s' \text{ if } \pi_n(i,s) \sim_n \pi_n(i,s') \text{ for all } n \in CClass \text{ and } i \in ID_n$$

where $\sim$ is the behavioural equivalence of the composed object, *CClass* is the set of classes of objects, $\sim_n$ is the behavioural equivalence for the class n, ID_n is the set of identifiers for the objects of class n, and π_n have the same meaning as in the static case.

If all projection operations are behavioural then the other implication holds too. ☐

For example, the behavioural equivalence for the telephone system is defined as

$$T_1 \sim T_2 \text{ if } \texttt{tel}(n, T_1) \texttt{ =*= } \texttt{tel}(n, T_2) \text{ for all } n \in \omega$$

and can be encoded in CafeOBJ by using a parameterized relation as follows:

```
mod COINDUCTION-REL {
  protecting(TELEPHONE-SYSTEM)

  op _R[_]_ : TelSys Number TelSys -> Bool {coherent}

  vars T1 T2 : TelSys
  var N : Number

  eq T1 R[N] T2 = tel(N, T1) =*= tel(N, T2) .
}
```

One may use this for proving, for example, the true concurrency of telephone calls as follows:

```
open COINDUCTION-REL .
  op telsys :    -> TelSys .
  ops n n1 n2 :    -> Number .
  ops u1 u2   :    -> Unit .
-- n =/= n1, n =/= n2
red call(n1, u1, call(n2, u2, telsys)) R[n]
    call(n2, u2, call(n1, u1, telsys)) .
-- n = n1 or n = n2
red call(n1, u1, call(n2, u2, telsys)) R[n1]
    call(n2, u2, call(n1, u1, telsys)) .
```

18   Applications in Rewriting Logic

In this section we present a couple of typical applications of rewriting logic in CafeOBJ.

Debt Systems

Consider the problem of simplifying a debt system between several agents by
mutual debt cancellation or by debt transfer. Each simplification step is meant
to reduce the total amount of debt in the system while preserving the individual balance of each agent. An atomic debt is represented as a triple A_1 n A_2
meaning that A_1 owes n currency units to A_2. A system of debts is a multi-set
of atomic debts (specified in CafeOBJ via an associative-commutative binary
operation).

```
mod! DEBT {
  protecting(INT)
  protecting(QID * { sort Id -> Agent })

  [ Debt < Debt* ]

  op ___ : Agent Nat Agent -> Debt
  op nil : -> Debt*
  op __ : Debt* Debt* -> Debt* {assoc comm}
}
```

Trivial debt cancellations or system simplifications are expressed as equations,
while more fundamental simplifications are expressed as transitions.

```
mod! DEBT-REDUCE {
  protecting(DEBT)

  var S : Debt*
  vars m n : Nat
  vars !A !B !C : Agent

  eq nil S = S .

  eq  !A 0 !B = nil .
  eq (!A m !B) (!A n !B) = !A (m + n) !B .
  eq !A m !A = nil .
```

```
    ctrans (!A m !B) (!B n !C) =>
           (!A m !C) (!B (n - m) !C)    if (m <= n) .
    ctrans (!A m !B) (!B n !C) =>
           (!A (m - n) !B) (!A n !C)    if (n <= m) .
}
```

The initial model of DEBT-REDUCE has equivalence classes (under associativity, commutativity, but also the other debt simplification related equations) of debt system configurations as elements, and system configuration simplifications (corresponding to the two transitions of the specification, but modulo all the equations of the specification) as transitions between these elements. The transitions encode a simplification algorithm; we suggest that the use of transitions for *declarative encoding of algorithms* is the most appropriate use of RWL in CafeOBJ.

The following two verifications (both of them giving true and both of them having a different normal form in the right hand side) show that the problem is *not* confluent.

```
open DEBT-REDUCE .
red ('c 3 'a) ('d 1 'a) ('a 2 'b) ==>
    ('d 1 'a) ('c 2 'b) ('c 1 'a) .
red ('c 3 'a) ('d 1 'a) ('a 2 'b) ==>
    ('c 2 'a) ('d 1 'b) ('c 1 'b) .
```

This basically means that equational logic is inappropriate for modeling this problem, since using only equations rather than transitions would identify fundamentally different configurations. On the other hand, rewriting logic gives the possibility to reason about configuration changes. For example one might define a function

```
    op balance : Agent Debt* -> Int
```

giving the individual balance for each agent, and prove a (inductive) property of invariance

```
    cq balance(a, sys1) == balance(a, sys2)
       if sys1 ==> sys2 .
```

or define a another function

```
op debt : Debt* -> Nat
```

giving the total debt in the system, and prove a simplification (inductive) property

```
cq debt(sys1) <= debt(sys2) if sys1 ==> sys2 .
```

Sorting Strings

Another simpler but more classical example is given by sorting algorithms. A crude version of the bubble sort algorithm might have the following very compact encoding in **CafeOBJ**:

```
mod! SORTING-STRG {
  protecting(STRG*(NAT)* { sort Nat -> Elt })

  ctrans N:Elt . N':Elt => N' . N  if (N' < N) .
}
```

where STRG* is the parameterized specification of strings of Example 22. This **CafeOBJ** specification is sorting strings of natural numbers. We can actually sort the strings by using the command execute as follows:

```
exec (4 . 3 . 5 . 3 . 1) .
```

and get

```
-- execute in SORTING-STRG : 4 . 3 . 5 . 3 . 1
1 . 3 . 3 . 4 . 5 : Strg
```

Notice that although the sorting problem is confluent, it is still important to specify it in rewriting logic rather than just equational logic since this enables reasoning about the algorithm. For example we can prove the termination of this algorithm by introducing a function measuring the "disorder degree" of a string:

```
mod! SORTING-STRG-PROOF {
  protecting(SORTING-STRG + NAT)

  op disorder : Strg -> Nat
  op _>>_ : Elt Strg -> Nat

  vars E E' : Elt
  var S : Strg

  eq disorder(nil) = 0 .
  eq disorder(E) = 0 .
  eq disorder(E . S) = disorder(S) + (E >> S) .

  eq E >> nil = 0 .
  cq E >> E' = 0 if E <= E' .
  cq E >> E' = 1 if E' < E .
  cq E >> (E' . S) = s(E >> S) if E' < E .
  cq E >> (E' . S) = (E >> S) if E <= E' .
}
```

The fact that the sorting algorithm always terminates can be then formulated
as

$$s \implies s' \text{ implies } \texttt{disorder}(s') < \texttt{disorder}(s)$$

19 General Small Methodological Advises

Finally, we give an ad-hoc list of small advises which we find very useful for
specification development and for the verification/testing stage.

1. Always write relatively small modules and small module expressions.

2. Small steps in the development of specifications should alternate with
   some proofs. The results of such proofs should be kept and used as
   lemmas for further proofs. In the case of complex systems, such gradual
   interleaving between the specification and the verification stages of the
   system development not only provides a healthy style of development,

but might also speed up the execution of the proof scores in the later stages of system verification.

3. Use coinduction rather than _=b=_ for behavioural properties proofs; _=b=_ should be used only for simple behavioural lemmas reductions.

4. Try to avoid non-*protecting* modes for imports and especially for parameterization. Protecting can be assured by avoiding the definition of new operations of imported sorts. When such operations are needed one may instead define and use supersorts of the imported sorts.

5. Try to visualize the module expressions developed by writing down their corresponding diagrams. Try to avoid module expressions "on the fly", better give names to the views and module expressions defined.

6. Switch easily from the verification to the testing stage of a specification by having the specification with only user defined data types but using views from them to the built-in data types.

7. Special attention should be paid to the new design of the modules system and especially to the adequate use of parameter sharing.

Bibliography

[1] Michael Barr and Charles Wells. *Category Theory for Computing Science*. Prentice-Hall, 1995.

[2] Garrett Birkhoff. On the structure of abstract algebras. *Proceedings of the Cambridge Philosophical Society*, 31:433–454, 1935.

[3] Rod Burstall and Răzvan Diaconescu. Hiding and behaviour: an institutional approach. In A. William Roscoe, editor, *A Classical Mind: Essays in Honour of C.A.R. Hoare*, pages 75–92. Prentice-Hall, 1994. Also in Technical Report ECS-LFCS-8892-253, Laboratory for Foundations of Computer Science, University of Edinburgh, 1992.

[4] Rod Burstall and Joseph Goguen. Putting theories together to make specifications. In Raj Reddy, editor, *Proceedings, Fifth International Joint Conference on Artificial Intelligence*, pages 1045–1058. Department of Computer Science, Carnegie-Mellon University, 1977.

[5] Rod Burstall and Joseph Goguen. The semantics of Clear, a specification language. In Dines Bjorner, editor, *Proceedings, 1979 Copenhagen Winter School on Abstract Software Specification*, pages 292–332. Springer, 1980. Lecture Notes in Computer Science, Volume 86; based on unpublished notes handed out at the Symposium on Algebra and Applications, Stefan Banach Center, Warsaw, Poland, 1978.

[6] Virgil Căzănescu. Local equational logic. In Zoltan Esik, editor, *Proceedings, 9th International Conference on Fundamentals of Computation Theory FCT'93*, pages 162–170. Springer-Verlag, 1993. Lecture Notes in Computer Science, Volume 710.

[7] Virgil Emil Căzănescu and Grigore Roşu. Weak inclusion systems. *Mathematical Structures in Computer Science*, 7(2), 1997.

[8] Răzvan Diaconescu. *Category-based Semantics for Equational and Constraint Logic Programming*. PhD thesis, University of Oxford, 1994.

[9] Răzvan Diaconescu. Completeness of category-based equational deduction. *Mathematical Structures in Computer Science*, 5(1):9–41, 1995.

[10] Răzvan Diaconescu. Behavioural rewriting logic: semantic foundations and proof theory, October 1996. Submitted to publication.

[11] Răzvan Diaconescu. A category-based equational logic semantics to constraint programming. In Magne Haveraaen, Olaf Owe, and Ole-Johan Dahl, editors, *Recent Trends in Data Type Specification*, volume 1130 of *Lecture Notes in Computer Science*, pages 200–221. Springer, 1996. Proceedings of 11th Workshop on Specification of Abstract Data Types. Oslo, Norway, September 1995.

[12] Răzvan Diaconescu. Constraint logics, December 1996. Submitted to publication.

[13] Răzvan Diaconescu. Foundations of behavioural specification in rewriting logic. In *Proceedings, First International Workshop on Rewriting Logic and its Applications. Asilomar, California, September 1996.*, volume 4 of *Electronic Notes in Theoretical Computer Science*. Elsevier Science, 1996.

[14] Răzvan Diaconescu. Extra theory morphisms for institutions: logical semantics for multi-paradigm languages. *Jour. of Applied Categorical Structures*, 1998. To appear; a preliminary version appeared as JAIST Technical Report IS-RR-97-0032F in 1997.

[15] Răzvan Diaconescu and Kokichi Futatsugi. Logical semantics for CafeOBJ. In *Precise Semantics for Modelling Techniques*, 1998. Proceedings of a workshop to be held in Kyoto, Japan. Preliminary version appeared as Technical Report IS-RR-96-0024S at Japan Advanced Institute for Science and Technology in 1996.

[16] Răzvan Diaconescu, Joseph Goguen, and Petros Stefaneas. Logical support for modularisation. In Gerard Huet and Gordon Plotkin, editors,

Logical Environments, pages 83–130. Cambridge, 1993. Proceedings of a Workshop held in Edinburgh, Scotland, May 1991.

[17] James Rumbaugh et al. *Object-Oriented Modeling and Design*. Prentice Hall, 1991.

[18] K. Futatsugi, J.A. Goguen, J. Meseguer, and K. Okada. Parameterized programming in OBJ2. In *Proc. of the 9th Intl. Conf. on Software Engineering*, pages 51–60, 1987.

[19] K. Futatsugi, J.A. Goguen, J. Meseguer, and K. Okada. Parameterized programming and its application fo rapid prototyping in OBJ2. In U. Ohno and Y. Matsumoto, editors, *Japanese Perspectives on Software Engineering*, pages 77–102. Adison-Wesley, 1989.

[20] Kokichi Futatsugi. Hierarchical software development in HISP. In T. Kitagawa, editor, *Computer Science and Technologies 1982*, pages 151–174. OHMSA/North Holland, 1982. Japan Annual Review in Electronics, Computer and Telecommunications Series.

[21] Kokichi Futatsugi. Trends in formal specification methods based on algebraic specification techniques — from abstract data types to software processes: A personal perspective. In *Proceedings, International Conference Commemorating the 30th Anniversary of the Information Processing Society of Japan*, pages 59–66. Information Processing Society of Japan, 1990.

[22] Kokichi Futatsugi, Joseph Goguen, Jean-Pierre Jouannaud, and Jose Meseguer. Principles of OBJ2. In *Proceedings of the 12th ACM Symposium on Principles of Programming Languages*, pages 52–66. ACM, 1985.

[23] Kokichi Futatsugi and Ataru Nakagawa. An overview of CAFE specification environment – an algebraic approach for creating, verifying, and maintaining formal specifications over networks –. In *Proc. 1st Intl. Conf. on Formal Engineering Methods*, pages 170–181. IEEE, 1997.

[24] Kokichi Futatsugi and Koji Okada. Specification writing as construction of hierarchically structured clusters of operators. In *Proceedings, 1980 IFIP Congress*, pages 287–292. IFIP, 1980.

[25] Kokichi Futatsugi and Koji Okada. A hierarchical structurting method for functional software systems. In *Proceedings, 6th International Conference on Software Engineering*, pages 393–402. IEEE, 1982.

[26] Kokichi Futatsugi and Toshimi Sawada. Design considerations for Cafe specification environment. In *Proc. OBJ2 10th Anniversary Workshop*, October 1995.

[27] Joseph Goguen. Memories of ADJ. *Bulletin of the European Association for Theoretical Computer Science*, 36:96–102, October 1989. Guest column in the 'Algebraic Specification Column.' Also in *Current Trends in Theoretical Computer Science: Essays and Tutorials*, World Scientific, 1993, pages 76–81.

[28] Joseph Goguen. *Theorem Proving and Algebra*. MIT, To appear.

[29] Joseph Goguen and Rod Burstall. Institutions: Abstract model theory for specification and programming. *Journal of the Association for Computing Machinery*, 39(1):95–146, January 1992.

[30] Joseph Goguen and Răzvan Diaconescu. An Oxford survey of order sorted algebra. *Mathematical Structures in Computer Science*, 4(4):363–392, 1994.

[31] Joseph Goguen and Răzvan Diaconescu. Towards an algebraic semantics for the object paradigm. In Harmut Ehrig and Fernando Orejas, editors, *Recent Trends in Data Type Specification*, volume 785 of *Lecture Notes in Computer Science*, pages 1–34. Springer, 1994.

[32] Joseph Goguen and Răzvan Diaconescu. An introduction to category-based equational logic. In V.S. Alagar and Maurice Nivat, editors, *Algebraic Methodology and Software Technology*, volume 936 of *Lecture Notes in Computer Science*, pages 91–126. Springer, 1995.

[33] Joseph Goguen and Grant Malcolm. A hidden agenda. Technical Report CS97-538, University of California at San Diego, 1997.

[34] Joseph Goguen and José Meseguer. Completeness of many-sorted equational logic. *Houston Journal of Mathematics*, 11(3):307–334, 1985.

[35] Joseph Goguen and José Meseguer. Eqlog: Equality, types, and generic modules for logic programming. In Douglas DeGroot and Gary Lindstrom, editors, *Logic Programming: Functions, Relations and Equations*, pages 295–363. Prentice-Hall, 1986. An earlier version appears in *Journal of Logic Programming*, Volume 1, Number 2, pages 179–210, September 1984.

[36] Joseph Goguen and José Meseguer. Models and equality for logical programming. In Hartmut Ehrig, Giorgio Levi, Robert Kowalski, and Ugo Montanari, editors, *Proceedings, 1987 TAPSOFT*, pages 1–22. Springer, 1987. Lecture Notes in Computer Science, Volume 250.

[37] Joseph Goguen and José Meseguer. Unifying functional, object-oriented and relational programming, with logical semantics. In Bruce Shriver and Peter Wegner, editors, *Research Directions in Object-Oriented Programming*, pages 417–477. MIT, 1987. Preliminary version in *SIGPLAN Notices*, Volume 21, Number 10, pages 153–162, October 1986.

[38] Joseph Goguen and José Meseguer. Order-sorted algebra I: Equational deduction for multiple inheritance, overloading, exceptions and partial operations. *Theoretical Computer Science*, 105(2):217–273, 1992. Also, Programming Research Group Technical Monograph PRG–80, Oxford University, December 1989.

[39] Joseph Goguen, James Thatcher, Eric Wagner, and Jesse Wright. Initial algebra semantics and continuous algebras. *Journal of the Association for Computing Machinery*, 24(1):68–95, January 1977. An early version is "Initial Algebra Semantics", with James Thatcher, IBM T.J. Watson Research Center, Report RC 4865, May 1974.

[40] Joseph Goguen, Timothy Winkler, José Meseguer, Kokichi Futatsugi, and Jean-Pierre Jouannaud. Introducing OBJ. In Joseph Goguen, editor, *Algebraic Specification with OBJ: An Introduction with Case Studies*. Cambridge. To appear.

[41] Anne Elisabeth Haxthausen and Friederike Nickl. Pushouts of order-sorted algebraic specifications. In *Algebraic Methodology and Software Technology*, volume 1101 of *Lecture Notes in Computer Science*, pages 132–147. Springer, 1996.

[42] Rolf Hennicker. Context induction: a proof principle for behavioural abstractions. In A. Miola, editor, *Proceedings, International Symposium on the Design and Implementation of Symbolic Computation Systems*, volume 429 of *Lecture Notes in Computer Science*, pages 101–110. Springer, 1990.

[43] Shusaku Iida, Kokichi Futatsugi, and Takuo Watanabe. Algebraic specification of distributed systems based on concurrent object-oriented modeling. In Elie Najm and Jean-Bernard Stefani, editors, *Formal Methods for Open Object-based Distributed Systems*. Chapman & Hall, 1996.

[44] Shusaku Iida, Michihiro Matsumoto, Răzvan Diaconescu, Kokichi Futatsugi, and Dorel Lucanu. Concurrent object composition in CafeOBJ. Technical Report IS-RR-98-0009S, Japan Advanced Institute for Science and Technology, 1998. Submitted to publication.

[45] Saunders Mac Lane. *Categories for the Working Mathematician*. Springer, 1971.

[46] F. William Lawvere. An elementary theory of the category of sets. *Proceedings, National Academy of Sciences, U.S.A.*, 52:1506–1511, 1964.

[47] P. Lincoln M. Clavel, S. Eker and J. Meseguer. Principles of Maude. In *Proceedings, First International Workshop on Rewriting Logic and its Applications. Asilomar, California, September 1996.*, volume 4 of *Electronic Notes in Theoretical Computer Science*. Elsevier Science, 1996.

[48] José Meseguer. Membership algebra as a logical framework for equational specification. Invited paper at the Workshop on Algebraic Development Techniques, Tarquinia 1997.

[49] José Meseguer. Conditional rewriting logic as a unified model of concurrency. *Theoretical Computer Science*, 96(1):73–155, 1992.

[50] José Meseguer. A logical theory of concurrent objects and its realization in the Maude language. In Peter Wegner Gul Agha and Aki Yonezawa, editors, *Research Directions in Object-Based Concurrency*. MIT, to appear 1993. Also, Technical Report SRI-CSL-92-08, July 1992.

[51] Ataru Nakagawa, Toshimi Sawada, and Kokichi Futatsugi. CafeOBJ user manual, 1997. Version 1.3, ftp://sra.co.jp/pub/lang/CafeOBJ/Manual/.

[52] Shin Nakajima and Kokichi Futatsugi. An object-oriented modeling method for algebraic specifications in CafeOBJ. In *Proc. 19th Intl. Conf. on Software Engineering*, pages 34–44. ACM, 1997.

[53] K. Ohmaki, K. Futatsugi, and K. Takahashi. A basic lotos simulator in OBJ. In *Proceedings of InfoJapan'90 Computer Conference, Part 1*, pages 497–504. IPSJ, 1990.

[54] K. Okada and K. Futatsugi. Supporting the formal description process for communication protocols by an algebraic specification language OBJ2. In *Proc. of Second International Symposium on Interoperable Information System (ISIIS'88), Tokyo*, pages 334–343, 1988.

[55] Benjamin Pierce. *Basic Category Theory for Computer Scientists*. Foundations of Computing Series. MIT Press, 1991.

[56] Andrew Stevens and Joseph Goguen. Mechanised theorem proving with 2OBJ: A tutorial introduction. Technical report, Programming Research Group, University of Oxford, 1993.

[57] Andrzej Tarlecki. On the existence of free models in abstract algebraic institutions. *Theoretical Computer Science*, 37:269–304, 1986. Preliminary version, University of Edinburgh, Computer Science Department, Report CSR-165-84, 1984.

[58] Andrzej Tarlecki. Moving between logical systems. In Magne Haveraaen, Olaf Owe, and Ole-Johan Dahl, editors, *Recent Trends in Data Type Specification*, Lecture Notes in Computer Science, pages 478–502. Springer, 1996. Proceedings of 11th Workshop on Specification of Abstract Data Types. Oslo, Norway, September 1995.

[59] Alfred Tarski. The semantic conception of truth. *Philos. Phenomenological Research*, 4:13–47, 1944.

[60] Martin Wirsing. Algebraic specification. In Jan van Leeuwen, editor, *Formal Models and Semantics*, volume B. MIT Press/Elsevier, 1990.

Appendix: CafeOBJ Syntax

This appendix (courtesy to Toshimi Sawada from Software Research Associates Inc., Japan) shows the syntax table of CafeOBJ and corresponds to the SRA implementation.

Syntax

We use an extended BNF grammar to define the CafeOBJ syntax. The general form of a production is

nonterminal ::= *alternative* | *alternative* | ⋯ | *alternative*

The following extensions are used:

a ⋯   a list of one or more *a*s.

a, ⋯   a list of one or more *a*s separated by commas: "a" or "a, a" or "a, a, a", etc.

{ a }   { and } are meta-syntactical brackets treating *a* as one syntactic category.

[a]   an optional *a*: " " or "a".

Nonterminal symbols appear in *italic* font. Terminal symbols appear in sans serif font: "terminal", and may be surrounded by " and " for emphasis or to avoid confusion with the meta characters used in the extended BNF. Terminal symbols other than self-terminating characters (see section "Self-terminating Characters") are referred in this document as *keyword*s.

CafeOBJ programs

program ::= { *module* | *view* | *eval* } ⋯

A CafeOBJ program is a sequence of *module* (module declaration), *view* (view declaration) or *eval* (*reduce* or *execute* term).

Module Declaration

module	::= *module_type module_name* [*parameters*]
	[*principal_sort*] "{" *module_elt* ··· "}"
module_type	::= module | module! | module*
module_name	::= *ident* _1
parameters	::= "(" *parameter,* ··· ")"
parameter	::= [protecting | extending]
	parameter_name :: *module_expr* _23
parameter_name	::= *ident*
principal_sort	::= principal-sort *sort_name*
module_elt	::= *sharing | import | sort | record | operation*
	| variable | axiom | comment _4
sharing	::= share "(" *module_name* ")"
import	::= { protecting | extending | using }
	"(" *module_name,* ··· ")"
sort	::= *visible_sort | hidden_sort*
visible_sort	::= "[" *sort_decl,* ··· "]"
hidden_sort	::= "*[" *sort_decl,* ··· "]*"
sort_decl	::= *sort_name* ··· [*supersorts* ···]
supersorts	::= < *sort_name* ···
sort_name	::= *sort_symbol*[*qualifier*] _5
sort_symbol	::= *ident*
qualifier	::= "."*module_expr*[*qualifier*]

---

[1] The nonterminal *ident* is for identifiers and will be defined in the Section "Identifiers".

[2] *module_expr* is defined in the Section "Module Expressions".

[3] If optional [protecting \| extending] is omitted, it is defaulted to protecting.

[4] *comment* is discussed in Section "Comments and Separators".

[5] There must not be any separators (see Section "Lexical Considerations") between *ident* and *qualifier*.

record	::=	record *sort_name* [*super* $\cdots$]
		"{" { *slot* | *comment* } $\cdots$ "}"
super	::=	"[" *sort_name* ["(" *slot_rename,* $\cdots$ ")"] "]"
slot	::=	*slot_name* : *sort_name*
		| *slot_name* = "(" *term* ")" : *sort_name*
slot_name	::=	*ident*
slot_rename	::=	*slot_name* -> *slot_name*
operation	::=	{ op | bop } *operation_symbol* :
		[*arity*] -> *coarity* [*op_attrs*] _ [6]
arity	::=	*sort_name* $\cdots$
coarity	::=	*sort_name*
op_attrs	::=	"{" *op_attr* $\cdots$ "}"
op_attr	::=	constr | associative | commutative
		| idempotent | { id: | idr: } "(" *term* ")"
		| coherent | strat: "(" *natural* $\cdots$ ")"
		| prec: *natural* | l-assoc | r-assoc _ [7]
variable	::=	var *var_name* : *sort_name*
		| vars *var_name* $\cdots$: *sort_name*
var_name	::=	*ident*
axiom	::=	*equation* | *cequation* | *transition* | *ctransition*
equation	::=	{ eq | beq } [*label*] *term* = *term* "."
cequation	::=	{ ceq | bceq } [*label*] *term* = *term* if *term* "."
transition	::=	{ trans | btrans } [*label*] *term* => *term* "."
ctransition	::=	{ ctrans | bctrans } [*label*] *term* => *term*
		if *term* "."
label	::=	"[" *ident* "]:"
comment	::=	[-- | --> | ** | **>] *ASCII character* $\cdots$

---

[6] *operation_symbol* is defined in Section "Operation Symbols".

[7] *natural* is a natural number written in arabic notation.

Module Expressions

```
module_expr ::= module_name | sum | rename | instantiation
                | "("module_expr")"
sum          ::= module_expr { + module_expr } ···
rename       ::= module_expr * "{"rename_map,···"}"
instantiation ::= module_expr "("{ ident[qualifier] <= aview}, ··· ")"
rename_map   ::= sort_map | op_map
sort_map     ::= { sort | hsort } sort_name -> ident
op_map       ::= { op | bop } op_name -> operator_symbol
op_name      ::= operation_symbol | "("operation_symbol")"qualifier
aview        ::= view_name | module_expr
                | view to module_expr "{"view_elt,···"}"
view_name    ::= ident
view_elt     ::= sort_map | op_view | variable
op_view      ::= op_map | term -> term
```

When a module expression is not fully parenthesized, the proper nesting of subexpressions may be ambiguous. The following precedence rule is used to resolve such ambiguities:

$$sum < rename < instantiation$$

View Declarations

```
view ::= view view_name from module_expr to module_expr
         "{" view_elt, ··· "}"
```

Evaluations

```
eval   ::= { reduce | behavioural-reduce | execute }
           context term "."
context ::= in module_expr :
```

The interpreter has a notion of *current module* which is specified by a *module_expr* and establishes a context. If this is set, *context* can be omitted.

Abbreviations

Module type

The following are abbreviations for *module_type*:

Keyword	Abbreviation
module	mod
module!	mod!
module*	mod*

Module Declarations

make ::= make *module_name* "(" *module_expr* ")"

make is a short hand for declaring a module of name *module_name* which imports *module_expr* in protecting mode.

Principal Sorts

principal-sort can be abbreviated to psort.

Importing Modes

For importing modes, the following abbreviations can be used:

Keyword	Abbreviation
protecting	pr
extending	ex
using	us

Simultaneous Operation Declarations

Several operations with the same arity, coarity and attributes can be declared at once by ops. The form

ops *operation_symbol*$_1$ $\cdots$ *operation_symbol*$_n$
: *arity* -> *coarity op_attrs*

is just equivalent to the following multiple operation declarations:

op *operation_symbol$_1$* : *arity* -> *coarity op_attrs*

$$\vdots$$

op *operation_symbol$_n$* : *arity* -> *coarity op_attrs*

> bops is the counterpart of ops for behavioural operations.

bops *operation_symbol* $\cdots$: *arity* -> *coarity op_attrs*

In simultaneous declarations, parentheses are sometimes necessary to separate operation symbols. This is always required if an operation symbol contains dots, blank characters, or underscores.

Predicates

Predicate declaration (*predicate*) is a syntactic sugar for declaring Bool-operations, and has the syntax:

predicate ::= pred *operation_symbol* : *arity* [*op_attrs*]   – [8]

The form

pred *operation_symbol* : *arity op_attrs*

is equivalent to:

op *operation_symbol* : *arity* -> Bool *op_attrs*

Operation Attributes

The following abbreviations are available:

Keyword	Abbreviation
associative	assoc
commutative	comm
idempotent	idem

---

[8]You cannot use *sort_name* of the same character sequence as that of any keywords, i.e., module, op, vars, etc. in *arity*.

Sentences

For the keywords introducing sentences, the following abbreviations can be used:

Keyword	Abbreviation	Keyword	Abbreviation
ceq	cq	bceq	bcq
trans	trns	ctrans	ctrns
btrans	btrns	bctrans	bctrns

Blocks of Declarations

References to (importations of) other modules, signature definitions and axioms can be clusterd in blocked declarations:

$imports$ ::= imports "{"
　　　　　　{ *import* | *comment* } ⋯
　　　　　　"}"
$signature$ ::= signature "{"
　　　　　　{ *sort* | *record* | *operation* | *comment* } ⋯
　　　　　　"}"
$axioms$ ::= axioms "{"
　　　　　　{ *variable* | *axiom* | *comment* } ⋯
　　　　　　"}"

Views

To reduce the complexity of views appearing in module instantiations, some sytactic sugars are provided.

Firstly, it is possible to identify parameters by positions, not by names. For example, if a parameterized module is declared like

```
module! FOO (A1 :: TH1, A2 :: TH2) { ... }
```

the form

```
FOO(V1, V2)
```

is equivalent to

```
FOO(A1 <= V1, A2 <= V2)
```

Secondly, view to construct in arguments of module instantiations can always be omitted. That is,

```
FOO(A1 <= view to module_expr{...})
```

can be written as

```
FOO(A1 <= module_expr{...})
```

Evaluations

Keyword	Abbreviation
reduce	red
behavioural-reduce	bred
execute	exec

Lexical Considerations

A CafeOBJ program is written as a sequence of tokens and separators. A *token* is a sequence of "printing" ASCII characters (octal 40 through 176).[9] A *separator* is a "blank" character (space, vertical tab, horizontal tab, carriage return, newline, form feed). In general, any number of separators may appear between tokens.

Reserved Words

There are *no* reserved words in CafeOBJ. One can use keywords such as module, op, var, or signature, etc. for identifiers or operation symbols.

Self-terminating Characters

The following seven characters are always treated as *self-terminating*, i.e., the character itself constructs a token.

$$( ) , [ ] \{ \}$$

---

[9]The current interpreter accepts EUC coded Japanese characters also, but this is beyond the definition of CafeOBJ language.

Identifiers

The nonterminal *ident* is for *identifiers* which is a sequence of any printing ASCII characters except the following:

> self-terminating characters
> . (dot)
> " (double quote)

Upper- and lowercase characters are distinguished within identifiers. *ident*s are used for module names (*module_name*), view names (*view_name*), parameter names (*parameter_name*), sort symbols (*sort_symbol*), variables(*var_name*), slot names (*slot_name*), and labels (*label*).

Operation Symbols

The nonterminal *operation_symbol* is used for naming operations (*operation*) and is a sequence of any ASCII characters (self-terminating characters or non-printing characters can be elements of operation names.)[10]

Underscores are specially treated when they apper as a part of operation names; they reserve the places where arguments of the operation are inserted. Thus the single underscore cannot be a name of an operation.

Separators

A *separator* is a blank character (space, vertical tab, horizontal tab, carriage return, newline, form feed). One or more separators must appear between any two adjacent non-self-terminating tokens.[11]

Comments also act as separators, but their apperance is limited to some specific places (see Section "Syntax").

---

[10]Except the EOT character (control-D).

[11]The same rule is applied to *term*. Further, if an *operation_symbol* contains blanks or self-terminating characters, it is sometimes neccessary to enclose a term with such operation as top by parentheses for disambiguation.

Appendix: Institutions

In this appendix we review some of the basic concepts and results on institutions, but also introduce some novel concepts dealing with the semantics of the multi-paradigm systems. A good introduction to institutions is [29], and [16] contains many results about institutions with direct application to modularization. The reference for the new mathematics for multi-paradigm systems is [14].

From a logic perspective, institutions are much more abstract than Tarski's model theory [59], and also have another basic ingredient, namely signatures and the possibility of translating sentences and models across signature morphisms. A special case of this translation is familiar in first order model theory: if $\Sigma \to \Sigma'$ is an inclusion of first order signatures and M is a Σ'-model, then we can form the *reduct* of M to Σ, denoted $M\!\restriction_\Sigma$. Similarly, if e is a Σ-sentence, we can always view it as a Σ'-sentence (but there is no standard notation for this). The key axiom, called the **satisfaction condition**, says that *truth is invariant under change of notation*, which is surely a very basic intuition for traditional logic.

Definition 52 An **institution** $\mathfrak{I} = (\mathbb{S}ign, Sen, \text{MOD}, \models)$ consists of

1. a category $\mathbb{S}ign$, whose objects are called **signatures**,

2. a functor $Sen\colon \mathbb{S}ign \to \mathbb{S}et$, giving for each signature a set whose elements are called **sentences** over that signature,

3. a functor $\text{MOD}\colon \mathbb{S}ign \to \mathbb{C}at^{op}$ giving for each signature Σ a category whose objects are called Σ-**models**, and whose arrows are called Σ-**(model) morphisms**, and

133

4. a relation $\models_\Sigma \subseteq |\text{MOD}(\Sigma)| \times Sen(\Sigma)$ for each $\Sigma \in |\mathbb{S}ign|$, called Σ-**satis-faction**,

such that for each morphism $\varphi \colon \Sigma \to \Sigma'$ in $\mathbb{S}ign$, the **satisfaction condition**

$$M' \models_{\Sigma'} Sen(\varphi)(e) \quad \text{iff} \quad \text{MOD}(\varphi)(M') \models_\Sigma e$$

holds for each $M' \in |\text{MOD}(\Sigma')|$ and $e \in Sen(\Sigma)$. We may denote the reduct functor $\text{MOD}(\varphi)$ by $\_\!\upharpoonright_\varphi$ and the sentence translation $Sen(\varphi)$ by $\varphi(\_)$. $\square$

The following table shows the software engineering meaning of institution concepts for the case of specification languages.

INSTITUTIONS	SPECIFICATION LANGUAGES
signatures	syntactic declarations in modules
sentences	axioms in modules
models	(possible) implementations of modules
model morphisms	refinement between implementations
satisfaction relation	the implementation satisfies the axioms of the module
signature morphism	module import
sentence translation	importing the module axioms
model reduct	restricting the implementation of the importing module to an implementation of the imported module

Definition 53 Let $\mathfrak{I} = (\mathbb{S}ign, Sen, \text{MOD}, \models)$ be an institution. For any signature Σ the closure of a set E of Σ-sentences is $E^\bullet = \{e \mid E \models_\Sigma e\}$[12]. (Σ, E) is a **theory** iff $E = E^\bullet$.

A **theory morphism** $\varphi \colon (\Sigma, E) \to (\Sigma', E')$ is a signature morphism $\varphi \colon \Sigma \to \Sigma'$ such that $\varphi(E) \subseteq E'$. Let $\mathbb{T}h(\mathfrak{I})$ denote the category of all theories in $\mathfrak{I}$, and $sign^{\mathfrak{I}}$ the forgetful functor $\mathbb{T}h(\mathfrak{I}) \to \mathbb{S}ign$. $\square$

For any institution $\mathfrak{I}$, the model functor MOD extends to $\mathbb{T}h(\mathfrak{I})$, by mapping a theory (Σ, E) to the full subcategory $\text{MOD}(\Sigma, E)$ of $\text{MOD}(\Sigma)$ formed by the Σ-models that satisfy E.

---

[12]Meaning that $M \models_\Sigma e$ for any Σ-model M that satisfies all sentences in E.

Theories and theory morphisms have the following meaning in specification languages:

INSTITUTIONS	SPECIFICATION LANGUAGES
theory	(flattened) module
theory morphism	module import, view, module parameter

Liberality is a desirable property expressing the possibility of free constructions generalizing the principle of "initial algebra semantics". General results [57] show that liberality is equivalent to the power of Horn axiomatizability.

Definition 54 A theory morphism $\varphi\colon (\Sigma, E) \to (\Sigma', E')$ is **liberal** iff the reduct functor $\_\!\restriction_\varphi\colon \text{MOD}(\Sigma', E') \to \text{MOD}(\Sigma, E)$ has a left-adjoint $(\_)^\varphi$.

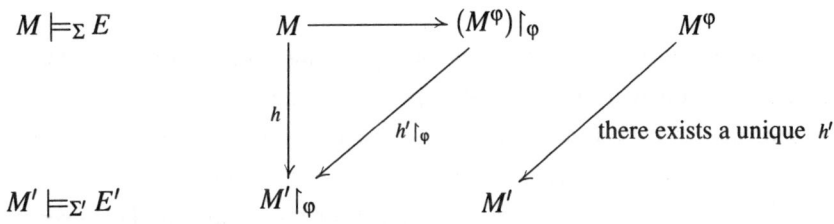

$$M \models_\Sigma E$$

$$M' \models_{\Sigma'} E'$$

The institution $\mathfrak{I}$ is **liberal** iff each theory morphism is liberal. When $\varphi\colon (\Sigma, \emptyset) \hookrightarrow (\Sigma, E)$, we denote $(\_)^\varphi$ by $\_/E'$. □

Another very important property expresses the possibility of amalgamation of consistent implementations for different modules (for more details see [16]); the following is its formulation within the framework of institutions:

Definition 55 An institution $\mathfrak{I}$ is **exact** iff the model functor $\text{MOD}\colon \mathbb{S}ign \to \mathbb{C}at^{op}$ preserves co-limits. $\mathfrak{I}$ is **semi-exact** iff MOD preserves only pushouts, and **weakly semi-exact** iff MOD maps pushouts to weak[13] pushouts. □

Semantics of multi-paradigm systems involves several different institutions which have to be linked together by using the following concept:

---

[13] See [45] for the definition and discussion of the concept of *weak* universal properties.

Definition 56 Let $\mathfrak{I}$ and $\mathfrak{I}'$ be institutions. Then an **institution morphism** $\mathfrak{I} \to \mathfrak{I}'$ consists of

1. a functor $\Phi\colon \mathbb{S}ign' \to \mathbb{S}ign$,

2. a natural transformation $\alpha\colon \Phi; Sen \Rightarrow Sen'$, and

3. a natural transformation $\beta\colon \text{MOD}' \Rightarrow \Phi; \text{MOD}$

such that the following **satisfaction condition** holds

$$M' \models_{\Sigma'} \alpha_{\Sigma'}(e) \text{ iff } \beta_{\Sigma'}(M') \models'_{\Phi(\Sigma')} e$$

for any Σ'-model M' from $\mathfrak{I}'$ and any $\Phi(\Sigma')$-sentence e from $\mathfrak{I}$. $\square$

In the literature there are several concepts of institution morphism, each of them being adequate to some specific class of problems. A good survey of various concepts of institution morphism discussing their usefulness can be found in [58]. The definition presented above and originally given by Goguen and Burstall [29] seems to be the most adequate for our approach. However, for obtaining some technical properties for extra theory morphisms, some technically stronger versions of this institution morphism are needed. These are very natural technical conditions which are easily satisfied in practice. The following definition is taken from [14]:

Definition 57 An institution morphism $(\Phi, \alpha, \beta)\colon \mathfrak{I} \to \mathfrak{I}'$ is

- **[strong] embedding** iff Φ admits a [left-inverse] left-adjoint $\overline{\Phi}$,

- **liberal** iff $\beta_{\Sigma'}$ has a left-adjoint $\overline{\beta}_{\Sigma'}$ for each $\Sigma' \in |\mathbb{S}ign'|$, and **persistent** iff in addition $\overline{\beta}_{\Sigma'}$ are left-inverses to $\beta_{\Sigma'}$ too, and

- **[weakly] additive** iff the squares defining the naturality of β are [weak] pullbacks.

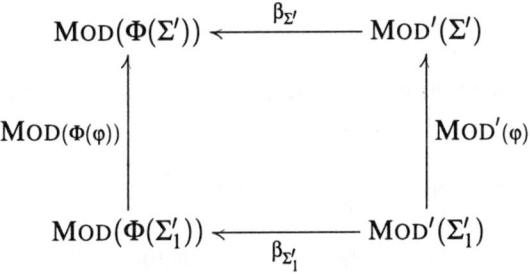

In the case of specification languages the components of institution embeddings have the following meaning:

INST.	SPECIFICATION LANGUAGES
Φ	reduces the syntax of modules to syntax in a simpler paradigm
$\overline{\Phi}$	regards the syntax of modules as (degenerated) syntax in a more complex paradigm
α	translates module axioms to axioms in a more complex paradigm
β	extracts a simpler paradigm implementation from a module implementation

Extra Theory Morphisms

Extra theory morphisms generalize the ordinary concept of theory morphism (Definition 53) in that it maps theories across an institution morphism. Intra (i.e., ordinary) theory morphisms can be regarded as special cases when the institution morphism is an identity. In this section we give a summary of the main definitions and results about extra theory morphisms supporting the semantics of structuring CafeOBJ specifications; all definitions and results of this section are taken from [14].

Definition 58 Let $(\Phi, \alpha, \beta) \colon \mathfrak{I} \to \mathfrak{I}'$ be an institution morphism, and $T = (\Sigma, E)$ and $T' = (\Sigma', E')$ be theories in $\mathfrak{I}$, and $\mathfrak{I}'$ respectively. A **extra the-**

ory morphism $T \to T'$ is an $\mathfrak{I}$-signature morphism $\varphi \colon \Sigma \to \Phi(\Sigma')$ such that $\alpha_{\Sigma'}(\varphi(E)) \subseteq E'$. $\square$

Fact 59 Any institution embedding $(\Phi, \alpha, \beta) \colon \mathfrak{I} \to \mathfrak{I}'$ gives rise to a functor $\Phi^* \colon \mathbb{T}h(\mathfrak{I}) \to \mathbb{T}h(\mathfrak{I}')$ defined by

$$\Phi^*(\Sigma, E) = (\overline{\Phi}(\Sigma), \alpha_{\overline{\Phi}(\Sigma)}(\zeta_\Sigma(E))^\bullet)$$

where ζ is the unit of the adjoint pair of functors $\Phi, \overline{\Phi}$. $\square$

Proposition 60 Let $(\Phi, \alpha, \beta) \colon \mathfrak{I} \to \mathfrak{I}'$ be an institution embedding and let $T \in |\mathbb{T}h(\mathfrak{I})|$ and $T' \in |\mathbb{T}h(\mathfrak{I}')|$. Then there is a natural bijection between extra theory morphisms $T \to T'$ and $\mathfrak{I}'$-theory morphisms $\Phi^*(T) \to T'$. $\square$

Model reducts are the semantic aspect of theory morphisms, therefore they play a central rôle in any semantics based on institutions. Model reducts for extra theory morphisms generalize ordinary model reducts for intra theory morphisms; they are introduced by the following result which can also be regarded as a Satisfaction Condition for extra theory morphisms:

Proposition 61 Let $(\Phi, \alpha, \beta) \colon \mathfrak{I} \to \mathfrak{I}'$ be an institution morphism. For any extra theory morphism $\varphi \colon (\Sigma, E) \to (\Sigma', E')$ there is a reduct functor $\_\!\upharpoonright_\varphi \colon \mathrm{MOD}(T') \to \mathrm{MOD}(T)$ defined by

$$M'\!\upharpoonright_\varphi = \beta_{\Sigma'}(M')\!\upharpoonright_\varphi$$

for M' any (Σ', E')-model. If (Φ, α, β) is an embedding, then

$$M'\!\upharpoonright_\varphi = \beta_{\Sigma\overline{\Phi}}(M'\!\upharpoonright_{\varphi'})\!\upharpoonright_{\zeta_\Sigma}$$

where $\varphi' \colon \overline{\Phi}(\Sigma) \to \Sigma'$ is the free extension of $\varphi \colon \Sigma \to \Phi(\Sigma')$. $\square$

The following extends the concept of liberality (free extensions) to extra theory morphisms:

Definition 62 A extra theory morphism $\varphi \colon (\Sigma, E) \to (\Sigma', E')$ is **liberal** iff the reduct functor $\_\!\upharpoonright_\varphi \colon \mathrm{MOD}'(\Sigma', E') \to \mathrm{MOD}(\Sigma, E)$ has a left adjoint, i.e., iff for any model $M \in |\mathrm{MOD}(\Sigma, E)|$, there exists a model $M' \in |\mathrm{MOD}'(\Sigma', E')|$ and a model morphism $\eta_M \colon M \to M' \!\upharpoonright_\varphi$ such that for any model $N' \in |\mathrm{MoD}'(\Sigma', E')|$ and any model morphism $h \colon M \to N' \!\upharpoonright_\varphi$ there exists a unique model morphism $h' \colon M' \to N'$ such that $h = \eta_M ; h' \!\upharpoonright_\varphi$.

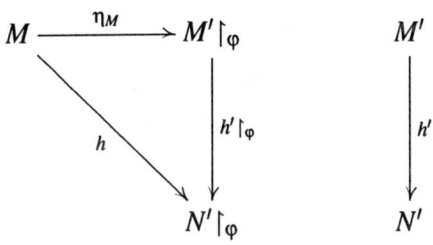

$\square$

The following results give sufficient conditions for "extra" liberality based on "intra" liberality.

Theorem 63 If $\mathfrak{I}$ is liberal on signature morphisms,[14] $\mathfrak{I}'$ and (Φ, α, β) are liberal, then any extra theory morphism $\varphi \colon (\Sigma, E) \to (\Sigma', E')$ is liberal. Moreover, the free (Σ', E')-model over a given (Σ, E)-model M can be obtained as $\overline{\beta}_{\Sigma'}(M^\varphi)/E'$. $\square$

Theorem 64 If (Φ, α, β) is a strong liberal embedding, $\mathfrak{I}'$ is liberal, and $\overline{\beta}$ satisfies the following Satisfaction Condition:

$$\overline{\beta}_{\Sigma'}(M) \models_{\Sigma'} \alpha_{\Sigma'}(e) \;\; \text{if} \;\; M \models_{\Phi(\Sigma')} e$$

for all $M \in |\mathrm{MoD}(\Phi(\Sigma'))|$ and $e \in Sen(\Phi(\Sigma'))$, then each extra theory morphism $\varphi \colon (\Sigma, E) \to (\Sigma', E')$ is liberal. $\square$

Corollary 65 If (Φ, α, β) is a persistent strong embedding and $\mathfrak{I}'$ is liberal, then any extra theory morphism is liberal. $\square$

The co-limit of a diagram of extra theory morphisms is computed in a pre-defined fixed institution in which all institutions underlying the nodes of the diagram are embedded. This is more general than just doing it in the co-limit of the underlying diagram of institution embeddings; this generality is

---

[14]I.e., The signature morphisms are liberal.

useful in applications since in the for multi-paradigm systems the co-cones of the underlying institutions are not necessarily co-limit co-cones. For details see [14], here we will mention only a corollary of the fundamental result on co-limits of theory morphisms:

Corollary 66 A diagram of extra theory morphisms has a co-limit whenever the co-limit co-cone of the underlying diagram of institution morphisms consists of institution embeddings. □

In general, extra theory morphisms do not necessarily support semi- exactness. Fortunately, some special cases of exactness for extra theory morphisms are enough to explain most practical situations. An important special case is given by the pushout between an intra and an extra theory morphism.

Theorem 67 Consider an institution embedding $(\Phi, \alpha, \beta) : \mathfrak{S} \to \mathfrak{S}_1$ and let $\varphi^{u2} : T \to T_2$ be a intra theory morphism in $\mathfrak{S}$, and $\varphi^{u1} : T \to T_1$ be a extra theory morphism with (Φ, α, β) the underlying institution morphism. If $\mathfrak{S}_1$ is (weakly) semi-exact and (Φ, α, β) is (weakly) additive and either of the following holds:

- (Φ, α, β) is strong, or

- $\mathfrak{S}$ is (weakly) semi-exact and Φ is surjective on objects and full

then the corresponding diagram of model reducts

$$
\begin{array}{ccc}
\mathrm{MOD}(T) & \xleftarrow{\ -\restriction_{\varphi^{u1}}\ } & \mathrm{MOD}^1(T_1) \\[1.5em]
\scriptstyle -\restriction_{\varphi^{u2}} \Big\uparrow & & \Big\uparrow \scriptstyle -\restriction_{\varphi^{1}} \\[1.5em]
\mathrm{MOD}(T_2) & \xleftarrow{\ -\restriction_{\varphi^{2}}\ } & \mathrm{MOD}^1(T_0)
\end{array}
$$

is a (weak) pullback. □

Inclusion systems where first introduced by [16] for the institution-independent study of structuring specifications. They provide the underlying mathematical concept for module imports, which are the most fundamental structuring construct. In this paper we use the *weak inclusion systems* of [7], which constitute a improvement of the original definition of inclusion systems of [16].

Definition 68 $\langle I, \mathcal{E} \rangle$ is a **weak inclusion system** for a category $\mathbb{C}$ if I and $\mathcal{E}$ are two sub-categories with $|I| = |\mathcal{E}| = |\mathbb{C}|$ such that

1. I is a partial order, and
2. every arrow f in $\mathbb{C}$ can be factored uniquely as $f = e; i$ with $e \in \mathcal{E}$ and $i \in I$.

The arrows of I are called **inclusions**, and the arrows of $\mathcal{E}$ are called **surjections**.[15] The domain (source) of the inclusion i in the factorization of f is called called the **image of** f and denoted as $\mathrm{Im}(f)$. An **injection** is a composite between an inclusion and an isomorphism. $\square$

For the fundamental properties of weak inclusion systems and techniques to construct them consult [7].

We need the following technical definition:

Definition 69 Let $\mathbb{C}$ and $\mathbb{C}'$ be two categories with weak inclusion systems $\langle I, \mathcal{E} \rangle$, and $\langle I', \mathcal{E}' \rangle$ respectively. Then a functor $\mathcal{U}: \mathbb{C} \to \mathbb{C}'$ **lifts inclusions uniquely** iff for any inclusion $\iota': A' \hookrightarrow B\mathcal{U}$ in I' with $B \in |\mathbb{C}|$, there exists a unique inclusion $\iota \in I$ such that $\iota\mathcal{U} = \iota'$. $\square$

Theorem 70 Consider a category of institutions with a weak inclusion system $\langle I^{\mathrm{INST}}, \mathcal{E}^{\mathrm{INST}} \rangle$ such that each of institutions involved $\Im = (\mathbb{S}ign, \mathrm{MOD}, Sen, \models)$ has a weak inclusion system $\langle I^{\Im}, \mathcal{E}^{\Im} \rangle$ for its category of signatures. If

- Φ preserves inclusions for each $(\Phi, \alpha, \beta) \in I^{\mathrm{INST}}$, and

---

[15]Surjections of some weak inclusion systems need not necessarily be surjective in the ordinary sense.

- Φ preserves both inclusions and surjections and lifts inclusions uniquely for each $(\Phi, \alpha, \beta) \in \mathcal{E}^{\text{INST}}$,

then the corresponding category of extra theory morphisms has an inclusion system where $\varphi : (\Sigma, E) \to (\Sigma', E')$ is

- *inclusion* iff both the underlying institution morphism
  $(\Phi, \alpha, \beta) : \mathfrak{I} \to \mathfrak{I}'$ and the signature morphism $\varphi : \Sigma \to \Sigma'\Phi$ are inclusions,
- *surjection* iff both the underlying institution morphism
  $(\Phi, \alpha, \beta) : \mathfrak{I} \to \mathfrak{I}'$ and the signature morphism $\varphi : \Sigma \to \Sigma'\Phi$ are surjections, and if $\alpha_{\Sigma'}(\varphi(E))^\bullet = E'$.

$\square$

Practical applications use mostly the following much simpler Corollary:

Corollary 71 Consider a partial ordered set of institutions and institution morphisms such that each of institutions involved
$\mathfrak{I} = (\mathbb{S}ign, \text{MOD}, Sen, \models)$ has a weak inclusion system $\langle I^{\mathfrak{I}}, \mathcal{E}^{\mathfrak{I}} \rangle$ for its category of signatures with Φ preserving inclusions for each institution morphism (Φ, α, β). Then the corresponding category of extra theory morphisms has an inclusion system where $\varphi : (\Sigma, E) \to (\Sigma', E')$ is

- *inclusion* iff the signature morphism $\varphi : \Sigma \to \Sigma'\Phi$ is an inclusion in $I^{\mathfrak{I}}$,
- *surjection* iff the underlying institution morphism is identity and it is a surjection in $\mathcal{E}^{\mathfrak{I}}$

$\square$

Appendix: a CASE study

In this appendix we present a CASE study which illustrates several aspects of specification and verification in CafeOBJ:

- behavioural specification and verification,
- specification code and behavioural equivalence reuse via the methodology of composing objects presented in this Report,
- module expressions (in a rather simple style), and
- automatic generation of analysis cases via meta-level encoding in CafeOBJ.

Many other aspects (most notably RWL) of this language are not present in an essential way in this example, this is bound to happen in most practical applications. We chose a behavioural specification example because this is one of the main distinctive novel features of this language, and because we believe that, due to its very efficient and economical abstraction mechanism, via adequate methodologies, behavioural specification emerges as the algebraic specification method of the future. We hope the reader will appreciate the clarity, simplicity, and high reusability (of both specification code and verification properties) of this specification style.

The CASE study presented is an (almost complete) specification of an ATM (i.e., *A*utomated *T*eller *M*achine) system and was developed as joint work with Shusaku Iida (whose specification writing style can be easily noticed) from JAIST.

The following OMT-like diagram gives the overall structure of the system:

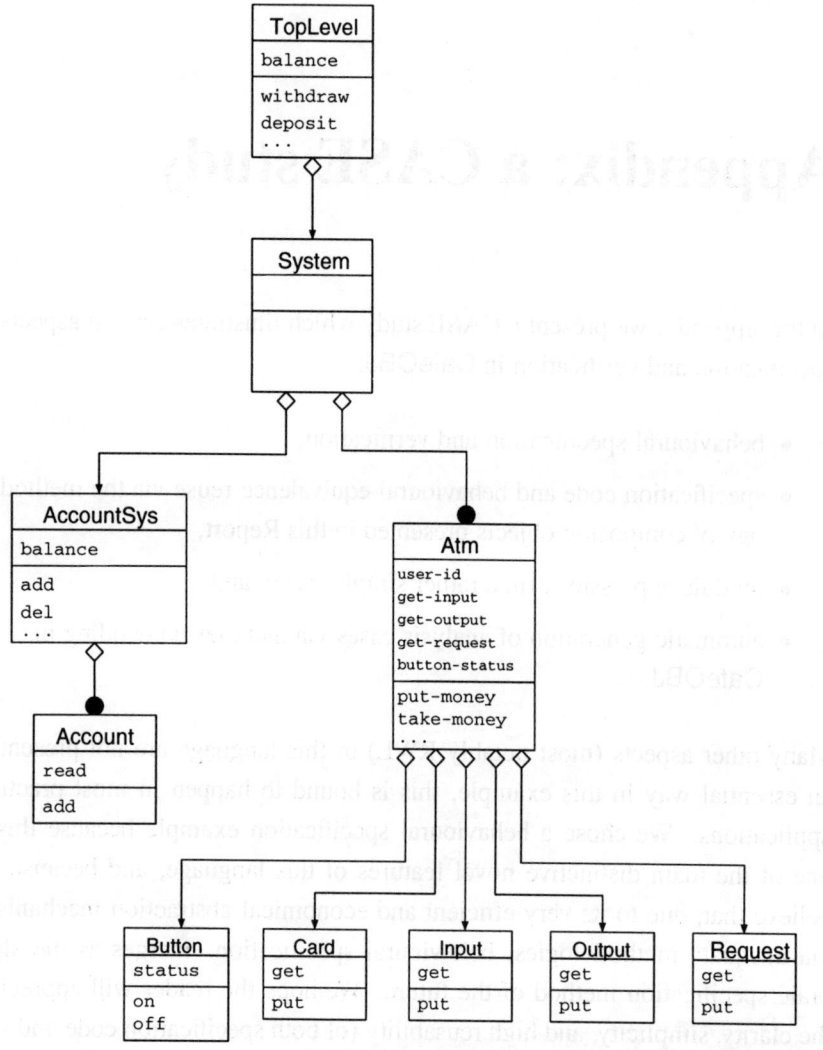

The whole system is a concurrent composition between a system of bank accounts and a dynamic system of ATM machines in a client-server relationship. An individual ATM machine is also a concurrent composition of several simple objects. The system of accounts is a dynamic composition of individual accounts.

We first specify a switch object with two values:

```
mod! ON-OFF {
  [ Value ]

  ops on off : -> Value
}

mod* SWITCH {
  protecting(ON-OFF)

  *[ Switch ]*

  -- initial state
  op init-sw : -> Switch
  -- switch on method
  bop on_ : Switch -> Switch
  -- switch off method
  bop off_ : Switch -> Switch
  -- state of the switch attribute
  bop status_ : Switch -> Value

  var S : Switch

  eq status(init-sw) = off .
  eq status(on(S)) = on .
  eq status(off(S)) = off .
}
```

Then we specify a counter object whose initialization is parameterized by user identifiers:

```
mod! USER-ID {
  protecting(NAT)
  [ Nat < UId ]

  op unidentified-user : -> UId
}

mod* COUNTER {
  protecting(USER-ID + INT)

  *[ Counter ]*
```

```
-- initialize counter with user ID
op init-counter : UId -> Counter
-- add a value to the counter (method)
bop add : Int Counter -> Counter
-- read the value of the counter (attribute)
bop read_ : Counter -> Int

var I : Int
var C : Counter
var U : UId

eq read(init-counter(U)) = 0 .
eq read(add(I, C)) = I + read(C) .
}

mod* COUNTER* {
  protecting(COUNTER)

  -- error value
  op counter-not-exist : -> Counter   -- error
}
```

The system of bank accounts is a dynamic concurrent composition of individual accounts, which are reused counters:

```
mod* ACCOUNT-SYSTEM {
  protecting(COUNTER*
     *{ hsort Counter -> Account,
        op init-counter -> init-account,
        op counter-not-exist -> no-account })

  *[ AccountSys ]*

  -- initial state
  op init-account-sys : -> AccountSys
  -- add a user account with user ID and
  -- initialize the deposit (method)
  bop add : UId Nat AccountSys -> AccountSys
  -- delete a user account (method)
  bop del : UId AccountSys -> AccountSys
  -- deposit method
  bop deposit : UId Nat AccountSys -> AccountSys
```

```
-- withdraw method
bop withdraw : UId Nat AccountSys -> AccountSys
-- balance of an user account (attribute)
bop balance : UId AccountSys -> Nat
-- get the state of a counter from
-- the state of an account (projection)
bop account : UId AccountSys -> Account

vars U U' : UId
var A : AccountSys
var N : Nat

eq account(U, init-account-sys) = no-account .
ceq account(U, add(U',N,A)) = add(N,init-account(U))
    if U == U' .
ceq account(U, add(U', N, A)) = account(U, A)
    if U =/= U' .
ceq account(U, del(U', A)) = no-account
    if U == U' .
ceq account(U, del(U', A)) = account(U, A)
    if U =/= U' .
ceq account(U,deposit(U',N,A)) = add(N,account(U,A))
    if U == U' .
ceq account(U, deposit(U', N, A)) = account(U, A)
    if U =/= U' .
ceq account(U,withdraw(U',N,A)) = add(-(N),account(U,A))
    if U == U' .
ceq account(U, withdraw(U', N, A)) = account(U, A)
    if U =/= U' .

eq balance(U, A) = read(account(U, A)) .
}
```

Next we define a simple cell object:

```
mod* TRIV+ {
  [ Elt ]

  op undefined : -> Elt
}

mod* CELL(X :: TRIV+) {
```

```
*[ Cell ]*

-- initial state
op init-cell : -> Cell
-- put the element to the cell (method)
bop put : Elt Cell -> Cell
-- get the element from the cell (method)
bop get : Cell -> Elt

var E : Elt
var C : Cell

eq get(init-cell) = undefined .
eq get(put(E, C)) = E .
}
```

The following are five simple specifications reusing the switch and the cell and for defining a button (for deposit and withdraw operations), a card slot, a withdraw request cell, and input and output cells:

```
mod* BUTTON {
  protecting(SWITCH *{ hsort Switch -> Button,
                       sort Value -> Operation,
                       op init-sw -> init-button,
                       op on -> deposit,
                       op off -> withdraw })
}
mod* CARD {
  protecting(CELL(X <= view to USER-ID
                  { sort Elt -> UId,
                    op undefined -> unidentified-user })
                 *{ hsort Cell -> Card,
                    op init-cell -> init-card })
}
mod* INPUT {
  protecting(CELL(X <= view to NAT
                  { sort Elt -> Nat,
                    op undefined -> 0 })
                 *{ hsort Cell -> Input,
                    op init-cell -> init-input })
}
mod* OUTPUT {
```

```
  protecting(CELL(X <= view to NAT
                      { sort Elt -> Nat,
                        op undefined -> 0 })
                  *{ hsort Cell -> Output,
                      op init-cell -> init-output })
}
mod* REQUEST {
  protecting(CELL(X <= view to NAT
                      { sort Elt -> Nat,
                        op undefined -> 0 })
                  *{ hsort Cell -> Request,
                      op init-cell -> init-request })

}
```

The ATM clients are concurrent composition of these small objects whose initialization is parameterized by ATM identifiers:

```
mod! ATM-ID {
  protecting(NAT *{ sort Nat -> AId })
}

mod* ATM-CLIENT {
-- importing data and the composing objects
  protecting(ATM-ID + BUTTON + CARD +
               INPUT  + OUTPUT + REQUEST)
  *[ Atm ]*

  -- initial state
  op init-atm : AId -> Atm
  -- errors
  op no-atm : -> Atm
  op invalid-operation : -> Atm
  -- push the deposit button (method)
  bop deposit : Atm -> Atm
  -- push the withdraw button (method)
  bop withdraw : Atm -> Atm
  -- input the request for withdraw (method)
  bop request : Nat Atm -> Atm
  -- put money method
  bop put-money : Nat Atm -> Atm
  -- take money method
  bop take-money : Atm -> Atm
```

```
-- set money for output (system operation; method)
bop set-money : Nat Atm -> Atm
-- put the bank card (method)
bop put-card : UId Atm -> Atm
-- clear all the info kept in the atm (method)
bop clear : Atm -> Atm
-- get the user ID (attribute)
bop user-id : Atm -> UId
-- get the inputed money (attribute)
bop get-input : Atm -> Nat
-- get the outputed money (attribute)
bop get-output : Atm -> Nat
-- get the request (attribute)
bop get-request : Atm -> Nat
-- get the state of the button (attribute)
bop button-status : Atm -> Operation

bop button : Atm -> Button        -- projection
bop card : Atm -> Card            -- projection
bop request : Atm -> Request      -- projection
bop input : Atm -> Input          -- projection
bop output : Atm -> Output        -- projection

var ATM : Atm
var N : Nat
var U : UId
var A : AId

eq button(init-atm(A)) = init-button .
eq button(invalid-operation) = init-button .
eq button(deposit(ATM)) = on(button(ATM)) .
eq button(withdraw(ATM)) = off(button(ATM)) .
eq button(request(N, ATM)) = button(ATM) .
eq button(put-money(N, ATM)) = button(ATM) .
eq button(take-money(ATM)) = button(ATM) .
eq button(set-money(N, ATM)) = button(ATM) .
eq button(put-card(U, ATM)) = button(ATM) .
eq button(clear(ATM)) = init-button .

eq card(init-atm(A)) = init-card .
eq card(invalid-operation) = init-card .
eq card(deposit(ATM)) = card(ATM) .
```

```
eq card(withdraw(ATM)) = card(ATM) .
eq card(request(N, ATM)) = card(ATM) .
eq card(put-money(N, ATM)) = card(ATM) .
eq card(take-money(ATM)) = card(ATM) .
eq card(set-money(N, ATM)) = card(ATM) .
eq card(put-card(U, ATM)) = put(U, card(ATM)) .
eq card(clear(ATM)) = init-card .

eq request(init-atm(A)) = init-request .
eq request(invalid-operation) = init-request .
eq request(deposit(ATM)) = request(ATM) .
eq request(withdraw(ATM)) = request(ATM) .
eq request(request(N, ATM)) = put(N, request(ATM)) .
eq request(put-money(N, ATM)) = request(ATM) .
eq request(take-money(ATM)) = request(ATM) .
eq request(set-money(N, ATM)) = request(ATM) .
eq request(put-card(U, ATM)) = request(ATM) .
eq request(clear(ATM)) = init-request .

eq input(init-atm(A)) = init-input .
eq input(invalid-operation) = init-input .
eq input(deposit(ATM)) = input(ATM) .
eq input(withdraw(ATM)) = input(ATM) .
eq input(request(N, ATM)) = input(ATM) .
eq input(put-money(N, ATM)) = put(N, input(ATM)) .
eq input(take-money(ATM)) = input(ATM) .
eq input(set-money(N, ATM)) = input(ATM) .
eq input(put-card(U, ATM)) = input(ATM) .
eq input(clear(ATM)) = init-input .

eq output(init-atm(A)) = init-output .
eq output(invalid-operation) = init-output .
eq output(deposit(ATM)) = output(ATM) .
eq output(withdraw(ATM)) = output(ATM) .
eq output(request(N, ATM)) = output(ATM) .
eq output(put-money(N, ATM)) = output(ATM) .
eq output(take-money(ATM)) = init-output .
eq output(set-money(N, ATM)) = put(N, output(ATM)) .
eq output(put-card(U, ATM)) = output(ATM) .
eq output(clear(ATM)) = output(ATM) .

eq user-id(ATM) = get(card(ATM)) .
```

```
   eq get-input(ATM) = get(input(ATM)) .
   eq get-output(ATM) = get(output(ATM)) .
   eq get-request(ATM) = get(request(ATM)) .
   eq button-status(ATM) = status(button(ATM)) .
}
```

The ATM system concurrently composes the dynamic system of ATM clients
with the system of bank accounts. Notice the rather heavy synchronization in
this composition:

```
mod* ATM-SYSTEM {
   protecting(ACCOUNT-SYSTEM + ATM-CLIENT)

   *[ System ]*

   -- initial state
   op init-sys : -> System
   -- add an atm to the system (method)
   bop add-atm : AId System -> System
   -- delete an atm from the system (method)
   bop del-atm : AId System -> System
   -- add an user account (method)
   bop add-user : UId Nat System -> System
   -- delete an user account (method)
   bop del-user : UId System -> System
   -- put the bank card (method)
   bop put-card : AId UId System -> System
   -- request for withdraw (method)
   bop request : AId Nat System -> System
   -- put money method
   bop put-money : AId Nat System -> System
   -- take money method
   bop take-money : AId System -> System
   -- deposit method
   bop deposit : AId System -> System
   -- withdraw method
   bop withdraw : AId System -> System
   -- push the OK button on atm to complete
   -- the operation (method)
   bop ok : AId System -> System
   -- cancel the operation of ATM (method)
   bop cancel : AId System -> System
```

```
-- get the balance of specified user (attribute)
bop balance : UId System -> Nat
-- projection operator for AccountSys
bop account-sys : System -> AccountSys
-- projection operator for Atm
bop atm : AId System -> Atm

var S : System
vars A A' : AId
var U : UId
var N : Nat

eq balance(U, S) = balance(U, account-sys(S)) .

eq account-sys(init-sys) = init-account-sys .
eq account-sys(add-atm(A, S)) = account-sys(S) .
eq account-sys(del-atm(A, S)) = account-sys(S) .
eq account-sys(add-user(U,N,S)) =
                            add(U,N,account-sys(S)) .
eq account-sys(del-user(U,S)) = del(U,account-sys(S)) .
eq account-sys(put-card(A, U, S)) = account-sys(S) .
eq account-sys(request(A, N, S)) = account-sys(S) .
eq account-sys(put-money(A, N, S)) = account-sys(S) .
eq account-sys(take-money(A, S)) = account-sys(S) .
eq account-sys(deposit(A, S)) = account-sys(S) .
eq account-sys(withdraw(A, S)) = account-sys(S) .
ceq account-sys(ok(A, S)) =
    deposit(user-id(atm(A, S)),
            get-input(atm(A, S)),
            account-sys(S))
    if button-status(atm(A, S)) == deposit and
       user-id(atm(A, S)) =/= unidentified-user and
       get-input(atm(A, S)) =/= 0 .
ceq account-sys(ok(A, S)) =
    withdraw(user-id(atm(A, S)),
            get-request(atm(A, S)),
            account-sys(S))
    if button-status(atm(A, S)) == withdraw and
       user-id(atm(A, S)) =/= unidentified-user and
       get-request(atm(A, S)) =/= 0 and
       get-request(atm(A, S)) <=
                balance(user-id(atm(A, S)),
```

```
                                          account-sys(S)) .
ceq account-sys(ok(A, S)) = account-sys(S)
     if user-id(atm(A, S)) == unidentified-user or
         (button-status(atm(A, S)) == deposit and
             get-input(atm(A, S)) == 0) or
         (button-status(atm(A, S)) == withdraw and
             (get-request(atm(A, S)) == 0 or
             get-request(atm(A, S)) >
                         balance(user-id(atm(A, S)),
                             account-sys(S)))) .
eq account-sys(cancel(A, S)) = account-sys(S) .

eq atm(A, init-sys) = no-atm .
ceq atm(A, add-atm(A', S)) = init-atm(A)
     if A == A' .
ceq atm(A, add-atm(A', S)) = atm(A, S)
     if A =/= A' .
ceq atm(A, del-atm(A', S)) = no-atm
     if A == A .
ceq atm(A, del-atm(A', S)) = atm(A, S)
     if A =/= A .
eq atm(A, add-user(U, N, S)) = atm(A, S) .
eq atm(A, del-user(U, S)) = atm(A, S) .
ceq atm(A, put-card(A', U, S)) = put-card(U, atm(A, S))
     if A == A' .
ceq atm(A, put-card(A', U, S)) = atm(A, S)
     if A =/= A' .
ceq atm(A, request(A', N, S)) = request(N, atm(A, S))
     if A == A' .
ceq atm(A, request(A', N, S)) = atm(A, S)
     if A =/= A' .
ceq atm(A, put-money(A',N,S)) = put-money(N, atm(A,S))
     if A == A' .
ceq atm(A, put-money(A', N, S)) = atm(A, S)
     if A =/= A' .
ceq atm(A, take-money(A', S)) = take-money(atm(A, S))
     if A == A' .
ceq atm(A, take-money(A', S)) = atm(A, S)
     if A =/= A' .
ceq atm(A, deposit(A', S)) = deposit(atm(A, S))
     if A == A' .
ceq atm(A, deposit(A', S)) = atm(A, S)
```

```
          if A =/= A' .
ceq atm(A, withdraw(A', S)) = withdraw(atm(A, S))
          if A == A' .
ceq atm(A, withdraw(A', S)) = atm(A, S)
          if A =/= A' .
ceq atm(A, ok(A', S)) = clear(atm(A, S))
          if A == A' and
             user-id(atm(A, S)) =/= unidentified-user and
             button-status(atm(A, S)) == deposit .
ceq atm(A,ok(A',S)) = set-money(get-request(atm(A,S)),
                                clear(atm(A, S)))
          if A == A' and
             user-id(atm(A, S)) =/= unidentified-user and
             button-status(atm(A, S)) ==  withdraw and
             get-request(atm(A, S)) <=
                     balance(user-id(atm(A, S)),
                             account-sys(S)) .
ceq atm(A, ok(A', S)) = invalid-operation
          if A == A' and
             (user-id(atm(A, S)) == unidentified-user or
             (button-status(atm(A, S)) ==  withdraw and
                (get-request(atm(A, S)) >
                     balance(user-id(atm(A, S)),
                             account-sys(S)))))) .
ceq atm(A, ok(A', S)) = atm(A, S)
          if A =/= A' .
ceq atm(A, cancel(A', S)) = init-atm(A)
          if A == A' .
ceq atm(A, cancel(A', S)) = atm(A, S)
          if A =/= A' .
}
```

Finally, we define some top level commands as abbreviations of complex sequences of method applications:

```
mod* ATM-SYSTEM-TOPLEVEL {
  protecting(ATM-SYSTEM)

  *[ TopLevel ]*

  -- initial state
  op init-tl : -> TopLevel
```

```
-- add a new atm (method)
bop add-atm : AId TopLevel -> TopLevel
-- delete an atm (method)
bop del-atm : AId TopLevel -> TopLevel
-- create an user account with initial balance
-- (method)
bop add-user : UId Nat TopLevel -> TopLevel
-- delete an user account (method)
bop del-user : UId TopLevel -> TopLevel
-- user "UId" goes to an ATM "AId" and deposits "Nat"
-- (method)
bop deposit : UId AId Nat TopLevel -> TopLevel
-- user "UId" goes to an ATM "AId" and withdraws "Nat"
-- (method)
bop withdraw : UId AId Nat TopLevel -> TopLevel
-- get a balance for the user (attribute)
bop balance : UId TopLevel -> Nat
-- projection operator for System
bop system : TopLevel -> System

var U : UId
var A : AId
var N : Nat
var TL : TopLevel

eq balance(U, TL) =
   balance(U, account-sys(system(TL))) .

eq system(init-tl) = init-sys .
eq system(add-atm(A, TL)) = add-atm(A, system(TL)) .
eq system(del-atm(A, TL)) = del-atm(A, system(TL)) .
eq system(add-user(U,N,TL)) = add-user(U,N,system(TL)) .
eq system(del-user(U, TL)) = del-user(U, system(TL)) .
eq system(deposit(U, A, N, TL)) =
   ok(A, put-money(A, N,
      deposit(A, put-card(A, U, system(TL)))))) .
eq system(withdraw(U, A, N, TL)) =
   take-money(A, ok(A, request(A, N, withdraw(A,
            put-card(A, U, system(TL)))))))) .
}
```

Until this stage, for the simplicity of presentation, we have not presented any proof scores or testing executions. But now, we may test this specification:

```
open ATM-SYSTEM-TOPLEVEL
   op tl : -> TopLevel .
   ops u1 u2 : -> UId .
   ops ai1 ai2 : -> AId .

red balance(u1, deposit(u1,ai2, 2,add-user(u1, 10,tl))) .

red balance(u1, withdraw(u1, ai1, 2, withdraw(u2, ai2, 3,
            add-user(u1, 10, add-user(u2, 10, tl))))) .
```

The first reduction evaluates the balance of the account of a user (u1) after opening an account with 10 and then deposit 2. The second reduction evaluates the balance of a user after two withdraw operations, one from the actual user and another from a different user. The result shows that the withdrawal of the other user does not affect the balance of the actual user. This last result shows a very basic safety property, but in this case this property is very particular.

The general formulation of such property is the concurrency of withdrawals by two different users without respect at which ATM machines these take place. This concurrency property has to be formulated as a behavioural equivalence property. The following module defines the behavioural equivalence for the whole system by using the results about the reusability of behavioural equivalence of the composing objects and the **CafeOBJ** default coinduction relation for the base level objects since for all these the default coinduction checking succeeds.

```
mod COINDUCTION-REL {
   protecting(ATM-SYSTEM-TOPLEVEL)
```

The following is the behavioural equivalence relation for ACCOUNT-SYSTEM which is parameterized by the user identifiers; and which reuses the behavioural equivalence on COUNTER* which is the default coinduction relation = * =.

```
   op _R[_]_ : AccountSys UId AccountSys -> Bool {coherent}
```

```
vars AS1 AS2 : AccountSys
var U : UId

eq AS1 R[U] AS2 = account(U, AS1) =*= account(U, AS2) .
```

The behavioural equivalence on the ATM-CLIENT is the conjunction of the behavioural equivalences of its composing objects; all of these are the default coinduction relations =*=.

```
op _R_ : Atm Atm -> Bool {coherent}

vars A1 A2 : Atm

eq A1 R A2 = button(A1) =*= button(A2) and
             card(A1) =*= card(A2) and
             request(A1) =*= request(A2) and
             input(A1) =*= input(A2) and
             output(A1) =*= output(A2) .
```

The behavioural equivalence for ATM-SYSTEM is the conjunction of the behavioural equivalences for ACCOUNT-SYSTEM and the conjunction of behavioural equivalences for all ATM clients (ATM-CLIENT).

```
op _R[_,_]_ : System UId AId System -> Bool {coherent}

vars S1 S2 : System
var A : AId

eq S1 R[U, A] S2 = account-sys(S1) R[U] account-sys(S2)
                   and  atm(A, S1) R    atm(A, S2) .
```

Finally, the behavioural equivalence at the top level is just the behavioural equivalence of the system.

```
op _R[_,_]_ : TopLevel UId AId TopLevel
                              -> Bool {coherent}

vars T1 T2 : TopLevel

eq T1 R[U, A] T2 = system(T1) R[U, A] system(T2) .
}
```

Now, we can proceed to do the proof of a behavioural property stating the true concurrency of cash withdrawals by different users without respect of the ATM machines involved or the amount of cash requested. At the top level this property can be expressed as

$$\texttt{withdraw}(u_1, A_1, N_1, \texttt{withdraw}(u_2, A_2, N_2, state)) \sim$$
$$\texttt{withdraw}(u_2, A_2, N_2, \texttt{withdraw}(u_1, A_1, N_1, state))$$

where u_i are users (identifiers), A_i are ATM machines (identifiers), and N_i are amounts of cash requested for withdrawal, for $i \in \{1,2\}$. Other parameters for this proof are the balance M_i of the accounts of the users u_i, for $i \in \{1,2\}$. The relationship between all these parameters amounts to a complex case analysis involving 108 cases. Fortunately, these can be automatically generated by **CafeOBJ** via a suitable meta-level encoding:

```
mod PROOF {
  protecting(COINDUCTION-REL)

  ops a a1 a2 : -> AId
  op t : -> TopLevel
  ops u u1 u2 : -> UId
  ops n1 n2 n01 n02 m1 m1' m2 m2' : -> Nat
```

Case analysis with respect to balances of accounts and requested amounts:

```
  eq n1 =/= 0 = true .
  eq n2 =/= 0 = true .
  eq n01 == 0 = true .
  eq n02 == 0 = true .
  eq n1 <= m1 = true .
  eq n01 <= m1 = true .
  eq n1 > m1' = true .
  eq n2 <= m2 = true .
  eq n02 <= m2 = true .
  eq n2 > m2' = true .
```

The following operations and equations generate the final proof term (RESULT)
which includes all cases generated by the case analysis:

```
op state-of-system : Nat Nat -> TopLevel
ops w1w2 w2w1 : AId AId Nat Nat Nat Nat -> TopLevel
op TERM : UId AId AId AId Nat Nat Nat Nat -> Bool
op TERM1 : UId AId AId AId Nat Nat -> Bool
op TERM2 : UId AId AId AId -> Bool
op TERM' : AId AId AId -> Bool
op RESULT :   -> Bool

vars A A1 A2 : AId
var U : UId
vars N1 N2 M1 M2 : Nat

eq state-of-system(M1, M2) = add-user(u1, M1,
                             add-user(u2, M2, t)) .

eq w1w2(A1, A2, N1, N2, M1, M2) =
   withdraw(u1, A1, N1,
   withdraw(u2, A2, N2, state-of-system(M1, M2))) .

eq w2w1(A1, A2, N1, N2, M1, M2) =
   withdraw(u2, A2, N2,
   withdraw(u1, A1, N1, state-of-system(M1, M2))) .
```

Notice that the balances of the accounts are specified via the method add-user;
this trick does not affect the generality of the proof.

The following sequence of equations gradually eliminates the parameters
by instantiating them to constants describing the case analysis:

```
eq TERM(U, A, A1, A2, N1, N2, M1, M2) =
   w1w2(A1,A2,N1,N2,M1,M2) R[U, A]
   w2w1(A1,A2,N1,N2,M1,M2) .
```

```
eq TERM1(U, A, A1, A2, N2, M2) =
      TERM(U, A, A1, A2, n1, N2, m1, M2) and
      TERM(U, A, A1, A2, n1, N2, m1', M2) and
      TERM(U, A, A1, A2, n01, N2, m1, M2) .

eq TERM2 (U, A, A1, A2) =
      TERM1(U, A, A1, A2, n2, m2) and
      TERM1(U, A, A1, A2, n2, m2') and
      TERM1(U, A, A1, A2, n02, m2) .

eq TERM'(A, A1, A2) = TERM2(u,  A, A1, A2) and
                      TERM2(u1, A, A1, A2) and
                      TERM2(u2, A, A1, A2) .
eq RESULT = TERM'(a,  a1, a2) and
            TERM'(a,  a1, a1) and
            TERM'(a,  a,  a) and
            TERM'(a1, a1, a2) .
```

The **CafeOBJ** system performs nearly 600,000 reductions for RESULT and gives true.

Notations and Conventions

In this Report we use the following notational conventions:

- **boldface** font for concepts and terminology introduced for the first time,

- *italic* font for ordinary emphasize, i.e., for concepts which were previously introduced or which are not especially defined in this Report,

- sanserif font for quotations of fundamental principles or rules,

- `teletype` font for CafeOBJ code or keywords, and

- **boldface sanserif** font for names of models.

The following table explains several mathematical notations used in this Report:

$\emptyset$	the empty set
$\|\mathbb{C}\|$	the class of objects of the category $\mathbb{C}$; the set of states in RWL
$\mathbb{C}^{op}$	the opposite of the category $\mathbb{C}$
$A \times B$	the products between the sets (or categories) A and B
$f: A \to B$	a function (or arrow) from A to B
$f;g$	the composition of functions (or arrows or transitions) f and g written in "diagramatic" order
A^B	the set of functions $f: B \to A$
$i: A \hookrightarrow B$	the inclusion of A into B seen as a function

$\eta: F \Rightarrow G$	natural transformation between the functors F and G
ω	the set of the natural numbers
$[n]$	the set $\{1,\ldots,n\}$ when n is a positive natural number, and $\emptyset$ when $n = 0$
$\trianglelefteq$	the module import relation
$\lhd$	the strict module import relation
$\trianglelefteq(pr)$	protecting import
$\trianglelefteq(ex)$	extending import
$SP(X :: P)$	parameterized specification (module)
$SP \wedge SP'$	the shared sub-modules of SP and SP'
$SP + SP'$	the shared sum of the specifications SP and SP'
Σ_{ws}	the set of operations of arity w and sort s, for the signature Σ
π	projection operation for object composition
(f,g)	signature morphism where f is the mapping of the sorts and g is the mapping of the operations
$\langle \Sigma, A \rangle$	built-in module, where Σ is the built-in signature, and A is the built-in model
M_s	the carrier of sort s of the model M
M_σ	the interpretation of the operation σ in the model M
$var(t)$	the set of variables in the term t
$(\forall X)\, t = t'$	(unconditional) equation
$(\forall X)\, t = t'$ **if** C	conditional equation
$(\forall X)\, t \Rightarrow t'$	(unconditional) transition
$(\forall X)\, t \Rightarrow t'$ **if** C	conditional transition
$(\forall X)\, t \rightarrow t'$ **if** C	TRS rule
$t \rightarrow t'$	1-step rewrite
$t \xrightarrow{*} t'$	arbitrary steps rewrite
$nf(t)$	the normal form of the term t
$(S,\Sigma,E),\ (\Sigma,E)$	equational specification (or theory)
(S,Σ,E,R)	RWL specification (or theory)
$(S,\leq,\Sigma)$	order sorted signature
$sign(SP)$	the signature of the specification SP

$0_{SP}, T_{SP}$	the initial model of the specification SP
T_Σ	the initial algebra of the signature Σ
$[\![SP]\!]$	the denotation of the specification SP
$\mathbb{S}et$	the category of sets and functions
$\mathbb{C}at$	the category of categories and functors
$\mathbb{S}ign$	the category of signatures for an institution
Sen	the sentence functor for an institution
MOD	the model functor for an institution
$\mathbb{T}h$	the theory functor for institutions
(Φ, α, β)	institution morphism
$M\!\restriction_\varphi$	the reduct of the model M through
	the signature morphism φ
$\models$	satisfaction relation
$\Vvdash$	behavioural satisfaction relation
$\vdash$	proof-theoretic consequence relation
$a \sim a'$	the behavioural equivalence relation
$\overline{M}$	the behavioural image of the model M

Index